SYSTEM CHANGE NOW!

"The prospects for humanity are somewhere between glorious and dire. It is hard to be more precise."

Colin Tudge *The Secret Life of Trees*

"To believe that doom is impossible ignores how terrible climate impacts could get.
But to believe that doom is inevitable ignores science, human ingenuity and the power of movements."

Lauren McLean on Twitter

"I believe in our human capacity to care deeply and act collectively! I believe in our ability to do what is right if we let ourselves feel it in our hearts.
So ... let us feel it in our hearts and act. We – the people with open hearts – will make the change"

Elizabeth Wathuti on Twitter 31/12/21

SYSTEM CHANGE NOW!

Richard Priestley

One man's vision of a better future
Healing a network of crises: Climate, Ecological, Political,
Social and Economic through a Global Green New Deal to achieve
Human Liberation & Planetary Rewilding

Castle Green Books 2022

Published 2022 by Castle Green Books
53 Mill Street Hereford HR1 2NX, UK

ISBN 978-1-3999-3540-1

Typeset and printed by
Short Run Press Ltd
Exeter UK

Contents

Chapter 6: Creating Change: What we can do

Chapter 7: Postscript: Putin & Possibilities

Foreword

This book, the movement and me

I am writing this book for anyone concerned about the many and diverse problems facing humanity, from the climate and ecological emergency to growing levels of inequality and poverty. It explores better systems of doing things and is written for anyone who looks at our current predicament and thinks, 'surely we could do better than this!' It is a book for people with a sense of curiosity who hope for a better future. It is explores solid examples of things that are changing for the better, and ideas about how things could be so very different, and far better still.

I write as an individual and I represent no organization. I'm now in my mid sixties and have been a climate, ecological and social activist for as long as I can remember. I gave my first talk linking all these three areas while still in Sixth Form, in 1972. Many previous drafts of this book have been produced and then abandoned. The time and energy I put into this was not wasted. People asked what I was writing about so in 2008 I started giving talks in pubs and church halls under the title 'Global Problems: Global Solutions'; then from 2010 I started a blog under the same title. Work started on this version of the book in tandem with a new series of online talks under the title 'Big Questions: Unthinkable Answers'.

This version has kept me busy from the autumn of 2020 to the spring of 2022. It has been a strange time with the Covid pandemic forcing many of us to re-evaluate how we as individuals live and interact with the world. Over the years I've been out on the streets on countless demonstrations, given talks at venues as varied as the Occupy protest camp by St Paul's Cathedral and in a committee room at the Houses of

Parliament, as well as doing quite a lot of leafleting and canvassing for the Green Party. Due to Covid and health concerns over this last year and a half I've largely been at home tending my vegetable garden and sitting at this computer, writing this book.

The ideas in this book represent a massive realignment in the global economy, and I make proposals requiring multi-trillion pound investments. My hope is that by talking about these unthinkable ideas I can in some way help shift the Overton Window and that enough people may feel inspired to turn some of these ideas into reality. I want to change the world. I can vote, go on demonstrations, get active within pressure groups, try to live an ethical lifestyle, but most powerful of all I hope are my words. I hope this book proves useful in our evolving shared response to the multiple crises we face. There are millions of us all pressing for the same kinds of system change.

As I write the COP26 climate conference in Glasgow is drawing to a close. The speeches inside the conference hall have largely been a 'festival of green-washing', in Greta Thunberg's memorable phrase. There have of course been notable exceptions, with serious and heartfelt calls for action by voices from the Global South, such as that made by the prime minister of Barbados Mia Mottley. The UK government is an extreme example of the disingenuous behaviour of most of the world leaders, saying fine words within the conference at the same time as enacting policies that show a total lack of commitment to the kinds of actions that absolutely must be made. The UK budget was presented to parliament the week before COP and it was a case study in exactly the wrong policies: reducing taxes on air travel, continuing with road and airport expansion, austerity for the poor and ever greater excess for the tiny mega-rich minority.

Outside the conference halls there have been huge demonstrations and eloquent speeches calling for real action with global equity and justice at the core of things. Greta Thunberg, Vanessa Nakate and many other leaders of the global climate movement have been prominent out on the streets of Glasgow. On 6th November there was a global day of action with demonstrations worldwide. Even here in Hereford we had over a hundred people gathered in High Town for an action that started and finished with singing and included fifteen minutes of silent vigil. For most of us here, and worldwide, it is the likes of Greta Thunberg and Vanessa Nakate whose views we respect, whereas Boris Johnson and Rishi Sunak represent the failed dominant ideologies that we hope to overturn.

Back in 2007 Paul Hawken wrote 'Blessed Unrest' in which he and a team of researchers tried to classify and calculate how many groups and organisations there were in the world actively campaigning for greater social and ecological justice. It was of course an almost impossible task, as groups grow and divide, some peter out and more are formed. He has since suggested the figure might be a million, but the number is not the important detail, it is what they can achieve that matters. They exist in pretty much every country on Earth and are linked up in the most amazing and decentralized way. Now in 2022 perhaps the best known groups are Extinction Rebellion and the School Strikes Movement but this can change quickly, and viewed from Brazil or India groups that are huge within those countries may receive very little in terms of global media coverage.

Many of us have been involved in more of these groups than we can even remember. It is not the individual groups that are important; it is their collective never ceasing energy and creativity, their striving to create something better that matters. During the eighteen months that I have been writing this book, peaceful demonstrations aimed at toppling governments have taken place in numerous countries. One extraordinary example has been Belarus. For about a year, from August 2020 until the summer of 2021 the vast majority of Belarusians have been supportive of, or actively participating in, wonderful and very well organised street protests, strikes and other non-violent actions, trying to topple the increasingly desperate authoritarian regime of Alexander Lukashenko. Around the world people know which side they are on. Despots and dictators naturally support Lukashenko. Ordinary people the world over support the heroic and peaceful protestors. What they want is pretty much what most people in most countries want: an end to the brutality and corruption and the opportunity to help create something better. What exactly that something better would mean would of course vary from person to person, but certainly be based on fairness and justice, economic security and opportunity, and the recovery of the natural world as pollution and destruction are curtailed.

As we saw with the Prague Spring in 1968 and the Arab Spring in 2010 the forces of repression can win. Now in December 2021 Lukashenko, with massive Russian support, looks to be crushing the pro-democracy movement. Success is never guaranteed, either for the forces of democratic progress or authoritarian repression. Lukashenko and Putin might well fall together, as might the brutal regimes of the Middle East. It might take some time. We need to think beyond the here and now and help imagine a different future, a better future. This seems a necessary first step and goes hand in hand with taking the kinds of action that may be of help to bring it into reality.

This book explores a vision of a future that might be appealing to people everywhere, and especially to the people who give me hope and inspiration now, and those who have done so throughout my lifetime. So I write for the amazing women leading the pro-democracy movement in Belarus. I write for the activists pressing for democratic governance in Sudan, and in many other countries. I write for climate activists everywhere, for the global school strikes movement and for Extinction Rebellion. I write for the poor and marginalized, and especially the poor of Africa who showed me such warmth and hospitality during my youthful travels. I write for the many authors who have provided me with inspiration over many years. I write for all those people who have come to my talks and who have given me feedback and nuggets of information that have helped shape my thinking and therefore the creation of this book. I write especially for the children, like my own grandchildren, born into the world at this time of great danger, and great opportunity. Their lives and their descendants' futures will be determined in large part by actions taken over the next decade.

Acknowledgements

Many people shaped my early thinking, noteworthy among them my grandparents Walter and Margaret Broughton, and my old geography teacher Tom Lancastle. Joseph Priestley, the eighteenth century scientist, philosopher and revolutionary, was my seven times great grandfather, and maybe the knowledge of his example somehow gave me permission to think in certain ways. Many speakers, writers and activists inspired me and perhaps none more so during the 1970s and 80s than Jonathon Porritt. A conversation in a pub with David Olivier in about 1996 led me into decades of researching renewable energy technologies.

Sadly my sister Liz Brady died before I started this version of the book. She had been enthusiastic about my ideas and encouraged me to write. Chris Cooke suggested I write a blog, and set one up for me and helped get me going as a blogger. I would like to thank Geoff Petty who made detailed comments on an earlier draft of this book. Jon Halle of Sharenergy, Mark E Thomas of the 99% Organization, Rob Garner of Herefordshire New Leaf Sustainable Development, Herbert Eppel of 100% Renewable UK and Perry Walker of Talk Shop have all helped and encouraged me over the years.

In the early stages of writing this version a small group of people made helpful suggestions, comments and feedback: thanks to David and Jane Straker, Beth Williamson, Patsy Wagner, Don Moules, Wendy Ogden, Nat Waring, Helen Lee and Tony Winch. Nigel Higgs, Temi Odurinde, Richard Urbanski, Rob Palgrave, Rob Hattersley and most of all Andrew Wibmer have all helped me with numerous computer related problems. For the last twenty-six years I have been in a men's group with Tristan Hill, Keith Cole, Claus Best, Jeff Beveridge, and the late David Gillett, and they have individually and collectively helped and supported me at a very profound level. My wife Colette has been my strong foundation, and has taught me so much about love and commitment.

Glenn Storhaug and Colette have between them tried their best to sort out my wonky spelling and grammar, and Glenn has proof read the text and prepared it for publication. I thank them all for their help. I'm sure that there are many other people I ought to thank who have in numerous ways over many years helped me in uncountable ways: thank you one and all.

I have done my best to fact check everything. There may well still be errors, and if so they are my responsibility.

Chapter One

System Change & A Global Green New Deal

'System Change, NOW!'

Climate activists chant 'System Change NOW!'
This is a book about system change. It proposes changes to pretty much everything that humanity is currently doing. These changes should not be seen as single stand-alone measures but as an integrated matrix of actions seeking to reverse an integrated matrix of crises.

The climate and ecological emergency is of course linked to other crises stemming from our current global systems of politics and economics, production and consumption. Conflict in Burkina Faso and growing economic insecurity for young people in Herefordshire may seem totally unrelated. At the heart of these and myriad other problems sits the dysfunction of our current system. That system needs to change.

There are 7.8 billion of us human beings on this precious and unique planet. There is also an abundance of financial and material resources. In order to meet the challenges ahead, all these human, material and financial resources need to be redeployed.

It is clear we have to get to net zero carbon emissions as fast as humanly possible. Extinction Rebellion is demanding this be done by 2025. Most governments are still talking about 2050 while carrying on with policies that will mean failure even to meet this ridiculously unambitious target. China has announced a target date of 2060 and India 2070. Most of these government plans rely on carbon capture technology that does not exist and will probably never exist at the required scale. The way governments measure carbon emissions exclude whole swathes of the economy from aircraft and shipping to the emissions of the military and of the embodied carbon emissions in imported goods. There is no sense of global social justice or intergenerational justice in anything stemming from almost all governments. Essentially, by setting these very long term goals, based on fantasy carbon accounting, with minimal short term action they are pushing the problems onto future generations. This is both morally indefensible and impractical. As the climate and biosphere break down the available windows of opportunity for action will close. It will simply be too late. Another of the climate protestors' banners I have read said 'Delayed action = death'. There is truth in that.

This book explores the many changes to politics and governance, tax and economics, energy generation and use, food and farming that may make extremely rapid decarbonisation achievable and in the process, socially beneficial.

Our current systems of farming and land use are destroying forests and soils, our vital carbon sinks and home to countless species of plant and animal life. The world's oceans are warming, acidifying, over-fished and full of plastic. All of this needs to change. We could feed all humanity a better diet while allowing nature to recover, but to do so the scale of change is immense. This book explores those changes and investigates how some of the best systems could be expanded to replace the worst, creating hundreds of millions of new jobs in the process.

Our current system favours the short term financial interests of mega rich individuals and corporations over the ecological stability of the planet and the economic security of the poor and marginalized. Austerity for the masses and excess for the tiny minority of the super rich have made many of our social ills very much worse. All this could be reversed. This book explores how.

Our systems of governance are utterly inadequate. They are over centralized, antiquated and corrupt. This book explores how some of the best models of decentralization and of global cooperation could be expanded and built upon to help bring into being all the other changes that are required to turn around the current crises. The laws of physics are immutable. The rules of political and economic life can and must be changed to reconcile the trajectory of the global economy with the health of the planet.

There is a whole world of ideas, projects, technologies and real world models of doing things that point the way to a better future where the climate and ecological emergency has been reversed and the social and political divisions that hamper human progress have been healed. Each of these things on their own cannot turn things around, but viewed as an integrated package and extrapolated into a global system of managing human affairs on this fragile planet, they do offer a pathway out of our current predicament.

Modern science is revealing our utter dependence upon myriad organisms of which until recently we were not aware. We could not breathe without the oxygen supplied for us by phytoplankton, or digest our food without the teeming multitudes of bacteria that live in our guts. We are not separate from nature but only live by our very embeddedness within it. This challenges the concept of dualism where mankind is seen as separate from and above nature, and a god perched above in some strange hierarchy. Modern science and ancient animism both point towards the vibrant, throbbing, interconnected biosphere as being our only true home, and heaven, and sometimes, especially if we make it so, hell.

At the very heart of any discussion of system change has to be some questioning of the concepts of economic growth and of capitalism. We need to negotiate our way into a post-capitalist future for which we do yet have a name or an adequate language to describe. We know that we cannot live without many things that economics cannot give a value to. We could not live without phytoplankton or bacteria or countless other species, and life without love, trust, friendship and solidarity would be no life at all, yet economics fails to give any of these things value. Meanwhile oil, coal, gas and thousands of commodities, technologies and infrastructure projects are given enormous value, when we could all live very well without them. Gross Domestic Product, GDP, measures how much money is spent in total within an economy and cannot distinguish between money spent on pointless wars or car crashes from money spent on useful things like food and health care. It is high time we created a new system. No one person is going to do this. No sudden and violent revolution could hope to bring it about. However the hive mind of humanity is collectively feeling its way into this new system. It is being created from below in a rapid process of grassroots, decentralized acts of social, political, philosophical and technological innovation, evolution and creation. This book is my contribution to that extraordinary and urgent act of collaborative creation.

Shifting the Overton Window

Most of these ideas, projects and developments that so inspire me are very little discussed in our media or by our politicians and are therefore outside the consciousness of most people most of the time. This is not surprising. Many of these ideas threaten the continued existence of a class of mega rich people and corporations with undue power and influence over our media and our politics. It is not in their interest that such ideas be discussed. These people shape the window of discourse, or Overton Window as it is sometimes known.

Extinction Rebellion exists to move the Overton Window. Their first demand is that governments and the media 'Tell the Truth' about the severity of the climate and ecological emergency. Without understanding the existential nature of the predicament we are in, governments and media remain blind to the possibilities of setting out on a dramatically different and very much better course.

This book and a series of online talks I'm giving called 'Big Questions: Unthinkable Answers' are my attempt to get people talking about both the scale of the problems and, perhaps even more importantly, the scale of the positive possibilities for change.

I draw the term 'unthinkable answers' from the concept of the Overton window. The Overton window is a term that refers to the range of policies politically acceptable to the mainstream population at any given time. It is also known as the window of discourse. Ideas are classed in six ranks, like planets in the solar system, from 'Policy', sitting in the centre of the Overton Window, out progressively through 'Popular', 'Sensible', 'Acceptable' to 'Radical' and finally to 'Unthinkable' in the far reaches of the Universe. Our politicians and media focus mainly on the policy and popular ideas, sometimes venturing out into the realms of the sensible and the acceptable, but almost never into the realms of the radical and even less so into the unthinkable. This is because their worldview is shaped by the predominant model of how the economy and politics work, and so they remain blind to those ideas that challenge that belief system, even when that belief system is leading to ever more extreme inequality, immense suffering, climatic chaos, and the extinction of countless species and the likely collapse of human civilization.

I want to give a real example of unthinkable ideas. About a decade ago I got my local MP Jesse Norman to introduce me to the people at Policy Exchange who were looking at carbon emissions and energy. All their research was going into carbon capture and storage and none into renewable energy. I put it to them that the world might move toward 100% renewables and that coal, oil and gas deposits would be left in the ground. To them such an idea was literally unthinkable. All their understanding of the economy was built on the assumption that fossil fuels were vital to the functioning of the economy. The idea that renewables might become cheaper to the extent that fossil fuels might become uncompetitive unexploitable stranded assets was unthinkable. They looked at me as if I was crazy. They had a big team of people at Policy Exchange and they all seemed to be employed to look at a problem with their pre-existing mindset preventing them from seeing the most significant changes in energy technology and energy costs. As for the wider implications of the whole climate and ecological emergency these of course were all well outside their consideration. They were generating vast amounts of data about issues that to me seemed peripheral and insignificant. They were missing important emerging trends and missing the big picture. This to me seems to typify our politics and media: their pre-existing understanding of the world prevents them from seeing the most important things. In terms of the Overton window they were firmly fixed in the zone of outdated policy, dismissing as crazy all the solutions that exist outside their little window, out in the world of ideas that to them were literally unthinkable.

Sometime later I discovered that the Policy Exchange gets funding from the fossil fuel industry, so in a sense it is hardly surprising that the idea that most fossil fuel deposits might switch from being resources with a vast economic value to being worthless stranded assets was unthinkable. So much of our current debate about the climate crisis is shaped by exactly this kind of blinkered thinking, deeply rooted in obsolete ideas about everything: technology, economics, politics and most of all about the health and wellbeing of our species and its place within the biosphere. So the ideas, technologies and policies that might be most useful to us are not seen and discussed; they remain unthinkable.

Jason Hickel's book 'Less is More: How Degrowth Will Save the World' explores how pervasive and destructive capitalism is with its endless need for economic growth on a finite and fragile planet. In the introduction he cites Fredric Jameson who once said that it is easier to imagine the end of the world than to imagine the end of capitalism. This seems to be the case. There are endless dystopian books and films about the future. Television and film exploration of a post capitalist world does not exist, as far as I am aware. Luckily books and blogs are a different matter, as ownership comes from below, whereas newspapers, TV and film are almost totally in the grip of the mega rich. We are so surrounded by global turbo-charged capitalism and the media embedded within it that it has become impossible to debate how we might transition out of this system and into another. Such debate is outside the Overton Window.

Governments, corporations and ordinary people all over the world need to face up to the reality of the climate and ecological emergency, and we all need to open up our imaginations and think about how we might set out on a very different course. We need to explore what system change might look like, and to do that the Overton Window needs to shift quite a long way. The 'unthinkable' has to become the new normal, the basis for policy decisions.

Of course many activists and greener politicians are talking about a change of direction. Recent youth-led demonstrations about climate change have featured the slogan 'Uproot the System'. Prominent among the emergent ideas is the concept of a Green New Deal. In the UK Caroline Lucas of the Green Party and in the USA Alexandria Ocasio-Cortez are leaders on this issue. These are necessary and good in themselves, but do not go far enough. In this book my task is to envision something very much more ambitious, a Global Green New Deal. Later in this chapter we will look at my outrageous proposals for a Global Green New Deal. Before that I want to step back and take a big-picture view of where we stand in the unfolding of human history, and how we may be on the verge of an epochal shift.

Epochal Shift

Future historians and archaeologists may write about the epochal shift from 'The Fossil Fuel Age' to 'The Solar Age'. Just as with previous periods like the Neolithic or the Bronze Age these ages may be defined in part by the tools and materials that were central to the people of those times. With each shift of technology there is often an associated and wide ranging shift in how people lived, what they ate and in their beliefs and values. People brought up in one era with its technologies and ideologies often struggle to hang on to that era in the face of a newly emerging set of ideas and technologies. The epochal shift may well be a time of confusion and conflict. We live in such times. It might be argued that we are witnessing the death throes of 'The Fossil Fuel Age'.

These future historians and archaeologists might characterise the 'Fossil Fuel Age' as having begun sometime around 1760 and finished around 2020. Starting in places like Ironbridge in Shropshire and Manchester a huge range of new technologies changed people's lives in many and diverse ways. This industrial revolution spread from Britain to Germany, USA and then to the vast majority of the world. Through empires, trade, colonialism and missionary activity there was nowhere beyond the reach of this new industrial, commercial and ideological paradigm. Huge demographic changes occurred as populations grew and cities sprung up on coalfields and as trading ports. Many of the changes from the pre-industrial world were good: medicine and sanitation cut disease and increased life expectancy; literacy, education and travel gave people many new opportunities and experiences. There were downsides: from industrial scale slavery and world wars to planetary-scale pollution and species extinction. Coal then oil and gas were the fuels that powered this era. From steam trains to jet planes, from the crockery we eat off to the clothes we wear, virtually everything in this era was in some way made by the use of fossil fuels.

Over these two-and-a-half centuries, from about 1760 to the present time, humanity's impact upon the Earth and every creature on it has been massively amplified. The balance of gases in the Earth's atmosphere, the acidity and temperature of the oceans, the structure and composition of soils, the number of species alive on the planet: all these things are now determined in large part by the actions of humans. Unwittingly we have become as gods. Our future and the future of very many other species rest upon our collective actions. In the words of Stewart Brand, 'we are as gods and HAVE to get good at it'. Scientists already refer to this new era as the Anthropocene. We have already entered the sixth great mass extinction event. Species are being made extinct in greater numbers and more quickly than at any time since the asteroid impact that killed the dinosaurs sixty-five million years ago. Climate Change and Ocean Acidification together have the capacity to endanger phytoplankton, on which the oxygen balance of the atmosphere in large part depends. Large mammals with big lungs, such as humans, are totally dependent upon having enough oxygen in the air: without phytoplankton our days would be numbered. Without bees to fertilize our crops and worms to build soil fertility starvation would be inevitable. Our utter dependence on a healthy biosphere cannot be overstated.

Throughout the industrial era there has been an evolving narrative critiquing the industrial paradigm. Some commentators have long assumed societal collapse a near inevitability, arguing that overpopulation and diminishing and scarce resources would lead to all-consuming levels of conflict. Oil and population have been two central areas of concern. Those most concerned about Peak Oil argued that without oil advanced human civilization would collapse. From Malthus to the Ehrlichs, people have argued that the carrying capacity of the

Earth was relatively fixed and that once population reached a certain level famine was inevitable.

I take the view that a cleantech revolution is now underway, showing that quitting oil, coal, gas, and indeed nuclear, is relatively straightforward. There is also sufficient evidence showing that global food production can be greatly increased and made much more ecologically sustainable. Stabilizing population growth at about 9 billion would be helpful, but we could in theory feed a lot more than that, if we get everything else right. Throughout this book I will use the term 'The Solar Age' to mean much more than a simple transition from how we use energy, or water, or grow food. As with previous epochal shifts there will be changes in how we live and travel, what we eat and most importantly in what we believe and how we interact as human beings, which will in turn transform the social, economic and political ways our species organizes itself. This book is not a critique of industrialism per se, but rather of the ways it was organized during 'The Fossil Fuel Age'. The challenge of 'The Solar Age' will be to build on the achievements of 'The Fossil Fuel Age' while repairing the damage done during that previous age.

Placing a single date for an epochal shift is seldom straightforward. Thomas Newcomen built his steam engine in 1712, Ironbridge was built in 1779, the first power loom was in 1785, the Liverpool to Manchester railway opened in 1830. Any of these dates might do as a signifier for the start of the Industrial Revolution, but 1760 is often taken as a nominal date. At the time these innovations were taking place few could have foreseen the decline of horse power or the many uses fossil fuels would be put to over the following couple of centuries. Likewise 'The Solar Age' might be thought of as beginning with some technical innovation. The French inventor Augustin Mouchot developed concentrating solar power driven steam engines in the 1860s which Frank Shuman built upon in his pioneering work in Egypt in 1912-13. The photovoltaic effect was demonstrated by Edmond Becquerel in 1839, but it took until 1954 for the first practical demonstration of a photovoltaic cell by Bell Laboratories. The first large scale solar power station was built in California in stages between 1984 and 1990 using concentrating solar thermal technology. Spain made great strides forward with this technology in the first decade of the new millennium. Photovoltaics and wind power also both really started to take off during this same decade. In December 2008 I had an article published in which I claimed that 'The Solar Age' had begun. The cost of photovoltaic panels has been falling for many years and is predicted to keep falling for years to come. Their falling price meant that in 2020 solar became the cheapest form of electricity in most countries of the world. It will be up to future historians to pick a date for the start of 'The Solar Age', if indeed they choose to use such a term.

I would define the technological characteristics of 'The Solar Age' as prioritizing pollution minimization, energy efficiency, the use of renewable sources of energy for electricity, heating and cooling, transport and industry, and the transition to a circular economy based on re-use of materials. Globally, digital technology has linked us together in ways previously unimaginable. We are connected up and communicating at huge speeds great volumes of data. New forms of democratic possibility are opening up. If we as a species can cooperate and use the amplified power of our new technology in sensible ways we can solve a great many of the problems facing humanity. There is a growing movement of people, represented in virtually every country on Earth, who are working toward the shared goals of social justice and ecological sustainability. Collectively they, or should I say we, are struggling to be the midwives of 'The Solar Age'.

This book is as much about the shifts in values and beliefs as it is about the change from fossil fuels to renewables. At the heart of this change is a greater emphasis on peace and cooperation. Nationalism and governments pursuing their own short term national self interest has been a central characteristic of 'The Fossil Fuel Age'. The importance of single powerful individuals deeply committed to promoting what they saw as their nation's interests was closely associated with this dying era. From Hitler and Stalin through Thatcher to Trump and Johnson, and the leaders of Europe's new far right, we see this line of the cult of the individual, of nationalism and a suspicion of 'foreigners'.

In earlier periods of history the nation was barely a concept. Through the Bronze Age and Iron Age personal identity was focused around the tribe, then followed by the Romans and imperial domination. Religion, royal and feudal power dominated the medieval period. Concepts of the nation state grew most powerfully through 'The Fossil Fuel Age'. It is my belief that we are witnessing a shift in many people's primary allegiance, away from tribe or nation towards species and planetary consciousness. Those who really understand the climate and ecological crisis know that we are truly all in this together. Any way forward has to embrace the rights of generations to come, the necessity of a fully functioning biosphere and the absolute need for human solidarity. On a dead planet there will be no nations, religions, economy or human life. Saving ourselves as a species requires of us a degree of cooperation and solidarity we have never yet achieved. If there is to be a human future it will have to be based on values, ideas and technologies unfamiliar to us steeped in the ways of 'The Fossil Fuel Age'.

Capitalism predates 'The Fossil Fuel Age' but has grown with it and facilitated its growth. Capitalism as a system requires endless economic growth, and that growth overwhelmingly stems from exploiting people and natural resources. The endless need for more resources, and the waste and pollution that stem from their use, is at the core of the climate and ecological crisis. Capitalism also requires ever more capital to invest and to make profits from. The less capitalism is regulated, the more profits go to an ever smaller but more powerful elite, while mass impoverishment awaits the vast majority of people. Inequality is one of the defining characteristics of unrestrained capitalism, and this has devastating impacts.

In the twentieth century violent revolutions took place with the goal of replacing capitalism with some kind of socialist system, but the resulting systems, at least in Russia and China, were just as ecologically destructive and socially brutal. More successful was the tempering of the worst aspects of capitalism by a process of reform. The Nordic region has been perhaps the most successful at sustaining these reforms, but there are many other examples of the successful rejection of the worst aspects of capitalism. In the USA Roosevelt's New Deal or in the UK Attlee's post-war government made great strides in the right direction only to be defeated by the resurgent forces of yet more aggressive capitalism in the years since Reagan and Thatcher.

We need to move into a post-capitalist era as a matter of urgency. It will not be achieved by violent revolution. One of the greatest things to emerge in the twenty-first century has been the very well organised, non-violent, direct action protest movement. It is a globally networked phenomenon, spanning the school climate strike activists, to Extinction Rebellion, the women led protest in Belarus and countless grassroots movements in almost every country on Earth. These people are seeking to overcome systems of exploitation and brutality, and they are doing it with love, commitment and determination. What they, or should I say we, achieve may well reverse the climate and ecological emergency, and that might necessarily entail ending the leadership of numerous autocrats, the many forms of inequality, and capitalism itself.

Modern turbo-charged capitalism and fossil fuel use have grown together into a destructive force of planetary proportions. Time is not on our side. Previous epochal shifts have tended to be relatively slow. The existential nature of the multiple crises we now face does not give us the luxury of a slow and careful transition. We need to exit 'The Fossil Fuel Age' with all the speed we can muster. The next decade or two will be a time of extraordinary change and it will be a bumpy ride.

There are forces with enormous power and huge vested interest in keeping to the existing status quo. Now, in 2022, these old forces look globally dominant, just as communist regimes did in Eastern Europe in 1988, or the French monarchy did in 1788. Sudden, dramatic change is likely, and necessary. The 2020s, 2030s and 2040s may well see the death throes either of turbo-charged fossil fuel driven capitalism, or of our species. It seems to me that the two cannot co-exist much longer. If our species does survive and thrive, it may look back on this critical decade, the 2020s, as the start of 'The Solar Age'.

*

Freedom & Human Liberation

For most of recorded human history people have been remarkably unfree. Under feudalism the vast majority of the people were peasants, beholden to powerful lords and landowners. Capitalism developed hand in glove with slavery. As slavery was abolished and capitalism grew ever more powerful the vast majority of humanity was caught up in a system that amounted for many people to wage slavery. Under capitalism the very concepts of freedom and liberty were co-opted to mean the economic freedom for already very rich individuals to exploit their workers and the natural resources of the Earth for their own individual and corporate profit and power. The increase in freedom of these few individuals was achieved at the expense of the vast majority of the other human beings who in many ways saw their own freedoms gradually eroded.

When the billionaire owned press rails against red tape, what is often meant by red tape is legislation to protect workers' rights and the environment. The attack on red tape is also usually associated with an attack on any form of taxation that impinges on the unlimited accumulation of wealth by the mega rich. When politicians follow the lobbying of these rich individuals and corporations and do indeed cut the proverbial red tape the result is usually falling wages and increased economic insecurity for workers, falling government funding for health, education and all the other things that make society function properly. Also this cutting of red tape means watered down protection of the environment, which inevitably means increases in pollution. The vast increase in sewage discharges into the rivers and coastal waters of the UK during 2021-22 has been a prime example of this cutting back of red tape and of corporate profiteering. The money that should have been invested in updating sewage systems was instead siphoned off to shareholder dividends and corporate tax havens. The economic freedom of the mega rich and of corporations has been achieved at the expense of the freedom the majority of citizens, who lose their freedom to live an economically safe and secure life, and their freedom to breathe clean air or to swim in clean rivers.

The very basis of freedom has to be re-examined. Most basic to the concept of freedom is the understanding that to be free people need to have a basic set of securities. They need to be free from anxiety and fear, free from hunger and homelessness, free from addiction and dependency. They need to have their basic needs met, and to live in safe and supportive communities.

Central to capitalism is the concept of economic scarcity. We are pitched in a continual battle of all against all, in a struggle for economic survival. Winner takes all and the devil takes the hindmost. In many pre-capitalist societies there was the perceived belief in the unlimited abundance of the natural world, and a belief that human dignity was based around sharing, generosity and on building good community cohesion. Travelling in Africa in the 1970s I was often flabbergasted by very poor people's outrageous hospitality and generosity. People who in some ways seemed materially poor often struck me as leading lives more obviously overflowing with joy and good humour, and with a richness of community life that we in the richer parts of the world had lost. Years later, studying social anthropology, I came to see how common this idea of abundance is, or was, in many pre-capitalist societies.

I do not want to portray these societies from the historical record and on the geographical fringes of the modern capitalist world as somehow a holy grail of perfection. Many problems existed. Access to education and to modern health care was limited and life expectancy was often short. When the rains failed people often went hungry. Inter-tribal conflict was common. Despite all of this, after my time in Africa I felt that there was a case for aid workers to be sent from Africa to help us in the West, with our problems associated with an atomized society where isolation, anxiety and depression are rife. We have much to learn from indigenous cultures, just as they have much to learn from us.

Could we take the best from both worlds? Could we provide a comfortable and secure life for all 7.8 billion people where everybody has access to excellent health care and education, and where also community solidarity is strong and the sounds of singing, of laugher and children playing replace the roar of traffic in our city streets? Where people take delight in sharing things and once again perceive the world as being a place abundant with all that they need to be happy? Yes, all this seems possible to me. All of this also seems absolutely necessary as we set in motion the kinds of policies needed to address the climate and ecological emergency. To achieve all of this we need to redistribute resources and re-define freedom. We all need fair and equal access to many kinds of freedom: the freedom from poverty and hunger, freedom from anxiety and depression, the freedom to breathe clean air and the freedom to have confidence in the long term survival of our grandchildren, their descendants, and of our species.

The economic freedoms of the tiny mega-rich elite cannot be allowed to trample on these bigger and more profound freedoms of the vast majority of humanity. The mega-rich and the media they control will squeal as their wealth is taken from them, as is proposed in the Global Green New Deal that I outline in this book. However they may be happier in the long run. Much research suggests that the extreme inequality that is integral to unrestrained capitalism is doing great psychological damage to the rich as well as the poor. Their greed keeps them fearful of the population at large, often isolated and living a kind of bunkered defensive lifestyle. Increasingly happiness and wellbeing are understood as being closely linked to strong community cohesion, where friendship, trust and connection with one's family, neighbours, community and society at large are strong. Yes, of course we need a minimum material and economic basis for happiness and wellbeing, but it is these social and psychological aspects that become ever more important once that basic level of economic security is reached.

As unrestrained capitalism inevitably leads to extreme inequality and to planetary scale pollution clearly it as a system must come to an end. There seem to be just two methods to do this: revolution or reform. From the late nineteenth century through the first three-quarters of the twentieth century violent revolution with the goal of bringing in communistic systems was perhaps the dominant ideological pathway. Overall these systems failed to produce outcomes that enhanced freedom and happiness. The communist party replaced the feudal lords but still people's lives were remarkably unfree. Back in the 1970s I lived in Berlin and visited East Berlin frequently. Over the following years I visited several communist countries. It was not all bad, but many aspects were. Generally people were not happy with what had been created, and I was visiting peaceful and relatively prosperous parts of the Eastern bloc. Communism was only achieved with much brutality and bloodshed, and in the end it simply didn't work.

There is a better way. The Nordic region provides a model. From being poor and unequal in the 1860s to being prosperous and equal by the 1960s a process of reform changed the region in many fundamental ways. The Nordic region also provides examples of many of the best environmental protections: still not good enough, but very much better than could ever be achieved in any country run along the lines of unrestricted capitalism. There are many other countries moving along similar pathways, towards fairer and more ecologically literate governance. They are a diverse bunch: Costa Rica, New Zealand, Uruguay, Chile, Bhutan.

In 'The Nordic Theory of Everything' the Finnish journalist Anu Partanen explores what she refers to as 'The Nordic Theory of Love'. I extrapolate from this towards a 'Global Theory of Love'. In subsequent chapters we will explore this in some detail, but for now it seems sufficient to say it provides a model for a global system that allows for a maximum of freedom for all, based on a strong social and economic system of safety and security, where equality and freedom are deeply embedded within the system. One could ask, is the Nordic region capitalist or socialist? Interestingly Anu Partanen hardly uses either of these terms, preferring to focus on the detailed ways in which life in the Nordic region is so very different from that in the USA. The Nordic region certainly has a more moderated and constrained form of capitalism than the USA, but maybe even this terminology misses the point. Partanen sees Nordic society structured around the welfare of all children and government's role as enabling the healthy development of all children. This implies a strong degree of social provision, of equality and increasingly of environmental protection. In this book we project these kinds of policies onto a global arena and further towards radically restorative policies for the natural world. Such a set of policies does not yet have a name. Post-capitalist, reformed capitalism, Green eco-socialist, the Scandinavian model or whatever, it might be up to future historians to give this new system a name. Naming things can obscure the reality, because people carry inside themselves diverse definitions of a single term. Understanding the detail, the outcomes and the general direction of societal evolution is what matters.

Human Liberation & Planetary Rewilding

Species are declining at an unprecedented rate. It is unquestionably in the interests of our own species to turn this situation around. We need to do a lot of rewilding. The natural world has the capacity to bounce back, if we manage to reverse the damage we are doing before whole ecosystems irrecoverably collapse. For decades nature reserves have been set up as if we could preserve them from the damage being done elsewhere. If the greatest dangers to our wildlife are those that are of planetary proportions, such as the heating of the world and the acidification of the oceans, then creating little reserves here and there is never going to be much good. We as a species need to stop emitting carbon dioxide and stop plastic pollution, stop deforestation and destructive infrastructure projects and systems of fishing and farming that are doing so much damage. This is of course all necessary and urgent. Rewilding is also necessary and urgent and this rewilding needs to be of planetary proportions. We need wildlife to recolonize our cities and our post industrial landscapes, our denuded agricultural landscapes and we need to protect our remaining ecological hotspots and carbon sinks.

We also need to rewild ourselves. We need to learn again our inseparable embeddedness within nature. We need to explore the role our species can play within the context of a well-functioning biosphere, with us as its custodian. This is a role that modern society has failed miserably at. We have much to learn from indigenous peoples, and from the various branches of modern science. Globally a vision of the kinds of actions that are required is emerging. There are many aspects of this that we will explore in later chapters.

Daily contact with the natural world is gradually being understood to be of crucial importance to individual happiness and wellbeing. This seems to have become a vital lifeline for many people during the Covid 19 pandemic. It is so important to have some aspect of the natural world within an easy walk of where we live. This can be street trees, gardens, parks, birdsong, woodlands or rivers or the ocean, they each have their own life, their own energy, with which we all relate in our own individual way. This frequent contact with the natural world should be seen as a fundamental human right, just as access to health care, education, economic security and all the other aspects of a well-regulated society should be.

The consumer-driven capitalism that has dominated much of the globe for the past sixty or seventy years has encouraged an endless striving after luxury. This felt need for comfort and luxury is an entrapment, a form of addiction that ties ever more people into ever more consumption. Oliver James's book 'Affluenza' caused quite a stir in 2007 when it detailed this process and the devastating psychological affects this has upon very rich individuals and on the whole of society.

The academic discipline of economics often refers to human needs as infinite. This is founded upon and closely tied to the concept that endless economic growth is necessary and a positive thing. As people become richer they will want ever more esoteric and expensive possessions and services. Tied in with this is the economic concept of scarcity. We are all in an endless battle for scarce resources, because if our needs are infinite then resources will by definition be inadequate to meet those needs. This sense of scarcity and the inadequacy of resources is heightened by the growing population and the finite nature of the single planet we all call home.

Within this competitive struggle to survive sits our perceived calibration of success as judged by what we consume. We are encouraged by advertising and the whole ethos behind consumer-driven capitalism to keep working, striving to do better, earn more and to buy higher status goods and services. But all of these goods and services have an ecological footprint. The planet cannot sustain so much consumption. And, in the end, we as individuals cannot sustain it either. This struggle for competitive consumption is fragmenting our societies and isolating us from each other and it is making us unhappy. If our individual needs are infinite how can we ever be satisfied? We will always be left feeling unfulfilled, which is the point. This is what keeps us hooked in to keep working, striving for more, and fuelling ever more consumption. There has to be another way to live and to structure our economy.

A new 'economics of enough' is emerging. As individuals we can step aside from consumer driven capitalism. We can decide that not owning a car is quite liberating. We can just buy less stuff. Clothing, electrical goods and furniture can often be had for next to nothing by buying second hand or happily receiving the old cast-offs from friends and family. Instead of endlessly going to garden centres we can make our own compost, save our own seed, take cuttings and divide plants, rather than needing to spend money buying these things. By growing our own fruit and vegetables we can save money. Saving money is helpful, but it is only a small part of the picture. There is a deep sense of satisfaction that comes from growing things, making things, mending things and repurposing things and that is important to us psychologically. By cutting back on our material needs we can free ourselves from those soul destroying jobs that most of us are or have been tied to for parts of our lives, or indeed for our entire working lives. In the pursuit of frugality and voluntary simplicity we can find a sense of personal liberation that can never be experienced in the endless cycle of infinite shopping and competitive consumption.

All sorts of powerful ideas are emerging in the fields of economics and politics and about how we can organize society. Taken together they represent an exciting alternative to our current planet-destroying system. At the heart of all these ideas is a new economic model that sees human material needs as limited, and that the resources of the world are rich, abundant and renewable if we use them wisely and share them fairly. From this flows a different view of human nature. The political ramifications of this are vast. Just how would global politics look if it was based on the fair sharing of resources? If we liberated all 7.8 billion of us from the chains of poverty and the chains of addictive and pointless consumption, the unleashing of human energy and potential would be tremendous. We could change the world. We could ensure our own collective survival as a species. Maybe we would ditch many outdated attachments to the things that divide us such as our divisions of class, gender, race, tribe, nationality or religious tradition, and focus instead on what unites us. The new goal would be our collective struggle to ensure the best future possible for all of our descendants, for the future of our species, and for the entire biosphere in which we are embedded. In our collective effort to build this better future we might find a personal sense of meaning, of satisfaction and purpose that consumer driven capitalism has failed to do. We just might be happier consuming less and sharing more. This of course is a philosophy as old as the hills. However the way it is manifesting in these turbulent times at the end of the 'Fossil Fuel Age' is interesting. It is at the core of the transition out of that era and into something new, the era I am calling 'The Solar Age'.

A Global Green New Deal: Context

I do not see change coming from above, or from a single overarching manifesto, political party or leader, or emanating from a single country. The most positive changes are coming from below, in a decentralized, emergent, eruption of intent, innovation and activism. The next section is entitled 'A Global Green New Deal'. It is my personal attempt to outline a kind of global policy wish list. What I write is not a prescription for top down policy as if I or any one person was going to enact such a policy, even if I would love to see all these policies enacted. Rather it is my contribution to an ongoing collective act of creation. In later parts of this book we shall look at numerous acts of political leadership on these issues, emerging in many ways in many places. My intention with this proposed Global Green New Deal, and with this book, is to contribute to that collective act of creation. The real world changes that inspire me are emerging piecemeal, but in a way that is loosely networked, so we learn from each other, and each incremental change can be built upon and improve what is happening elsewhere. It may be that the United Nations, ramping up on the ambition of the Paris Agreement, ends up advocating this as a set of policies. Leadership will likely come from groups like the School Strikes Movement, Extinction Rebellion and the many newly emerging pro-democracy movements in places like Belarus, and from countless writers and doers. Where I do not expect leadership to come from are the governments of the biggest countries, or the biggest corporations, or mega-rich individuals. They are all too deeply embedded in the current system founded upon greed and exploitation. However, elections can change things and produce unexpectedly good outcomes.

A Global Green New Deal: The Proposal

The objectives of this Global Green New Deal would be multiple:

1. To achieve net zero carbon emissions, globally and extremely rapidly.
2. To allow nature to heal and for biodiversity to flourish.
3. To end, or minimize, all forms of pollution.
4. To end poverty and inequality and the many associated social problems.
5. To end, or massively reduce, war and conflict.
6. To provide everyone on Earth with good food, housing and energy.
7. To provide everyone on Earth with free, excellent education and health care
8. To provide everyone on Earth with worthwhile work and economic security.
9. To improve physical and mental health, purposefulness and happiness.
10. To achieve a well-functioning, participatory, socially just, global democracy.

This may seem a rather over ambitious agenda. The problems facing humanity are many and complex and all inter-related. The solutions form a similar matrix. We will not reverse the climate and ecological emergency in the time available without system change. This system change requires the complete re-orientation of the global economy. There is a lot of work to be done. A Global Green New Deal should provide a guarantee of meaningful and worthwhile work for everyone who wants to participate. What follows is an outline of some tax and spending proposals on a truly gigantic scale. Global GDP is currently about $88 trillion per year. I will outline proposals for spending a very large percentage of this figure, but first we need to look at how we might stop spending money in ways that are causing the damage, and also at some new forms of taxation that might be of use both to raise funds and suppress the economic logic that is fuelling the destruction of our world.

Raising the Funds
Slashing suicidal subsidies & spending

Fossil fuels are still being subsidized on a huge scale, by 5.2 trillion dollars per year or 6.5% of global GDP in 2017.[1] By 2020 they have risen to $5.9 trillion.[2] Perverse subsidies are still going into systems of farming and fishing that are destroying the soils, seabed and fish stocks. All these subsidies should be scrapped. Huge investments are being made into inappropriate infrastructure projects such as airport expansion and road building, the HS2 rail project and Hinkley C nuclear power station. Around the world there are thousands of these inappropriate infrastructure projects. Every year many trillions of dollars, pounds or Euros could be saved by not investing in these kinds of things. Global military spending in 2019 was $1.917 trillion.[3] This should be cut to zero, or pretty near zero (although this can probably only be implemented after many other changes have been made).

Over the thirteen years from the global financial crash of 2008 to 2021, governments and central banks have pumped $25 trillion[4] into the global economy under a policy called quantitative easing. Essentially it is money injected into the economy to keep the wheels of capitalism turning. Every single penny of this money could have been spent very differently, helping establish an ecologically sustainable and socially just global economic system.

Tax

Pigovian taxes are taxes on activities that have negative side effects on society. By taxing such activities it makes them less profitable and widespread and consequently less damaging, while simultaneously raising useful funds. A carbon tax is the best known example. A global carbon tax of at least £100 per tonne should be introduced immediately, and applied globally. It may be advantageous for a clear path of annual increases to be set out, so people and corporations can plan their exit strategies from fossil fuels.

All forms of mining should be taxed in order to make materials recycling more economic and wastefulness of resources less economic. The nuclear industry has always been subsidized, and the long term waste storage at public expense, so a tax on nuclear fuel rods should be introduced as it has in Germany. Taxes designed to dampen down demand and reduce the material throughput of the economy should be introduced. A tax on all forms of advertising would be very useful and reduce the harmful psychological effects of being constantly bombarded by advertising and also reduce the huge flows of material going to landfill. Sales taxes, particularly on luxury goods, could also be introduced or increased.

Taxing the rich

The latest report from Oxfam shows how the richest 1% of people are responsible for double the carbon emissions of the world's poorest 50% of people.[5] The world simply does not need billionaires, or the mega rich. They should be taxed out of existence. Top rates of income tax should be increased, probably to 100%, so nobody could earn more than a certain global limit. Corporation tax should also be dramatically increased. A Tobin or Robin Hood tax on financial transactions should be introduced, and taxes on wealth, land, property and inheritance should all be increased.

Currently the very rich individuals and corporations use tax havens and tax loopholes to evade tax. Nation states seem powerless to control such greed, and governments such as the current UK one seem positively to revel in it, as it makes them and their key funders ever richer, at everybody else's expense. A global tax authority with the power to close any tax haven or tax loophole should be created. Tax evasion should be treated as a crime of greater severity than international terrorism: it certainly is responsible for more deaths. The Pigovian taxes on everything from carbon to advertising should all of course be globally coordinated and administered. Countries competing to be low tax locations have led to austerity for the poor and excess for the rich, and to the lowering of environmental standards just at a time when they should have been radically improved.

A Global Green New Deal: Investing for People & Planet

Between 1966 and 1980 the world cooperated and eradicated smallpox. This is perhaps humanity's greatest achievement to date. In our current crisis our need for global co-operation has never been greater.

Decarbonising the global economy will require cooperation and investment on an unprecedented scale. It affords us the possibility to end poverty, ignorance and much preventable suffering. We should invest in nine main areas:

1. Renewable energy systems to supply 100% of humanity's energy needs, including for electricity, heating, cooling, industry and transport.
2. The establishment of a circular economy utilizing the very best cleantech to minimize pollution, materials throughput and energy consumption.
3. A global healthcare system, excellent quality and free at point of use to all 7.8 billion people wherever they happen to be.
4. A global education system, from all ages from Kindergarten to post-graduate, and including apprenticeships. Again, free and excellent.
5. Massive changes to how we manage the land, rivers and seas, ensuring very much more ecologically sustainable and socially

just use of these precious resources.

6. Democratic renewal: decentralised, active and participatory local government, linked together in networks and with an enhanced and directly elected United Nations.

7. Global legal rights to a home, with electricity, clean water and sewage systems, a right to health care and to education, to work and training, a legal right to clean air and water and access to land and to nature. A right to economic security, including a Global Universal Basic Income.

8. Active peace-building, mediation and the creation of new economic opportunities that make peaceful cooperation more economically viable and socially beneficial.

9. A Global Trust for People and Planet to help retrain and redeploy the global workforce into projects that help ensure the above eight points are achieved.

The main body of this book explores each of these nine areas of investment and how they might change lives and livelihoods in several specific locations around the world, from a farm in Herefordshire to the city of Ouarzazate in Morocco.

Tilting the Field

Individually and collectively we tend do what is cheapest and easiest. Under our current system this often means it is the most environmentally destructive and socially exploitative thing that usually gets done, even when there are alternatives that would be very much kinder to people and planet. The better alternatives exist, but are on the very margins of people's consciousness. Ideas and technologies fail to get financial backing and so are never developed although it would be wonderful if they did. The raft of taxes and subsidies that are at the core of my proposed Global Green New Deal are all aimed at tilting the field, so doing what is right for the planet and for people becomes the thing that actually gets done. This applies to every sector of the economy.

Currently fast fashion generates a lot of very difficult to recycle clothing that nearly all ends up in landfill. Applying the logic of a circular economy, clothes would be made to be durable, leased rather than purchased and the manufacturer would have responsibility for full end of use recycling. This would then determine the materials used in manufacture in the first place, as ease of re-use would become an important economic driver. Sales taxes and taxes on fossil fuels and on many materials, coupled with taxes on advertising would all serve to push up prices and drive down demand. Subsidies could additionally be used to kick start clothes leasing, materials recycling and other aspects of a circular economy.

The energy economy has never internalized the externalities of production. The full range of pollution (from climate disrupting carbon dioxide emissions to diesel particulates clogging the lungs of our children to oil spills wreaking havoc on our marine ecosystems) has never been a factor in determining the price we pay for a litre of petrol or diesel. So petrol and diesel has been far too cheap and easy for much too long. The taxes proposed here would make it a very much more expensive option. Investing heavily in walking, cycling and public transport, and rethinking urban design, work and shopping habits, would all help tilt the field to make doing what is good for human health and wellbeing, and eliminating or reducing pollution , the cheapest and easiest thing to do.

Likewise in the food and farming sector: soils have been degraded, forests destroyed and watercourses polluted, because the system that farmers operated under didn't reward their conservation. That needs to change. There are many ways in which we could produce more and better food without the fossil fuels, herbicides, pesticides, or even the artificial fertilizers, upon which modern agriculture has become dependent. Plenty of alternatives exist. Currently they are often more expensive and therefore restricted to niche markets, or not developed at all beyond the odd experimental project.

Across every sector reversing this economic logic is at the heart of my tax and investment proposals. We have to make what is most ecologically restorative and sustainable, and what is most socially beneficial, what actually gets done. It has to become the cheapest and easiest option. We shall explore all these better ways of doing things in subsequent chapters.

A Global Trust for People and Planet

I have explored how money could be saved by cutting out many forms of spending, and more money generated through massively increased taxation focused on polluting practices and excessively rich individuals and corporations. This might raise tens of trillions of dollars, pounds or Euros. Say we had an annual sum of $44 trillion, or 50% of current global GDP: how best to invest it to meet our long list of goals?

A universal basic income of one dollar per day paid to all 7.8 billion humans would only cost $2.8 trillion per year. This would have a dramatically positive effect on the lives of the poorest billion or so people. It would be of little help to the poor in richer countries where the cost of living is often very much higher. Local rates of UBI might be introduced locally. However, having an initial global system paying everyone exactly the same amount would be symbolically good in making manifest ideas of human equality.

Although UBI is effective in lifting people out of poverty and is in itself a good thing, there is another way of creating change that I want to focus upon. There is so much that needs to be done, and much of it is not now, and may never be, profitable. Therefore these things do not get developed at sufficient scale and speed, even when they are absolutely vital for our future flourishing as a species. We need to create a workforce to do these things. I see some kind of Global Trust for People and Planet having a key role in this. It would act as a kind of global agency offering education, training and employment geared toward very specific goals. It would work with all levels of government, with businesses and other agencies to create the workforce and projects to do what needs to be done to turn the global situation around. We will look in more detail at this organisation and the kinds of possible projects it might lead on in more detail later.

Mark Z Jacobson calculates that the cost of transitioning the entire global economy from fossil fuels to renewables would be about $73 trillion, to include all sectors and all countries. For this to be achieved within a decade an annual budget of $7.3 trillion would seem about right. However, as money is saved in diminishing health and social costs due to pollution and climate change, and avoided fuel costs in terms of coal, oil and gas not purchased, it has been calculated that this whole energy transition should soon pay for itself.

A global health and care system might require an annual budget of perhaps ten trillion dollars and a global system of education, from nursery to post graduate and including apprenticeships and training, perhaps another ten trillion per year. Ecological restoration and regenerative farming would also require some major investment of perhaps five trillion dollars per year. Better systems of governance, especially at the local and the global scale, would also require considerable investment. This again might pay for itself, as spending by the governments of nation states could be correspondingly reduced.

Achieving all these many and huge objectives cannot be left to the market, or to the governments of nation states who have continually prioritised their own narrow self interest over that of the wider community. I am proposing both stronger global governance and radical decentralization, and the overwhelming share of our multi-trillion dollar budget should be raised and invested at these levels. In subsequent chapters we shall look at how an interlinked network of projects might be used as a vehicle for this investment. We in the UK are currently used to seeing the ownership of infrastructure, businesses and services either in the hands of large corporations, or as nationalized industries. In other places, notably Germany and the Nordic region, co-operatives and municipalities own, control and provide services, as they did in Victorian and Edwardian Britain to a greater extent than now. The co-operative and municipal model has many advantages, and some ideas of vastly expanding this sector will be explored later.

The Global Trust for People and Planet is the organization I see as providing the workforce to achieve the vast array of highly ambitious goals we have set ourselves under this envisaged Global Green New Deal. It would be the organization channelling funds down from global tax raising authorities into a vast number of innovative projects in local communities all around the world, spawning millions of new businesses, co-operatives and municipal organisations employing hundreds of millions of people.

The following chapter looks at governance in some detail and proposes some quite radical changes. This will help set the context for our Global Green New Deal investment projects and the work of the 'Global Trust for People and Planet'.

Chapter Two

System Change: Politics, Economics & Society

Part One: Politics

What People Want

Most people want the same basic things. Peace, safety and security, food and shelter, good health care and education, meaningful and fulfilling work and strong ties with family, friends and community. They also want a government that is not corrupt and a general sense of fairness. Increasingly a healthy environment is being seen as absolutely fundamental. People want a stable climate, clean rivers and oceans, the knowledge that creatures they may never see such as tigers and polar bears are doing well, and they also want individual access to the healing power of the natural world close to where they live. Perhaps most importantly of all, at a very deep level, they want to have solid grounds for hope, the hope for a safe and secure future for themselves, their children, grandchildren and their descendants.

Governments vary a lot, but none can provide all of these things in isolation. Young people are terrified to have children and fearful for their own futures.[6] They see the pathetic lack of action from governments to avert climatic and ecological catastrophe and they are rightly furious. In many countries inequality is getting worse and governments seem incapable of reining in the infinite greed and recklessness of a class of very rich individuals and corporations that evade their responsibilities and their taxes. In such countries a process of mass impoverishment is going on while a tiny class of mega rich are getting vastly richer. Such extreme inequality is a recipe for social division and anger. Many governments are corrupt, many are brutal and authoritarian. People are fed up with them: they want real democracy, transparency and fairness.

More and more people are joining a global array of non-violent direct action protest movements determined to bring about system change, and this necessarily means sweeping away many existing governments.

One of the goals of this book is to try and define what unites these many protest movements and to understand the kind of system change that they might all like. To do that we need to envisage a different system from any that currently exists. Theoretically, providing humanity with the things we all need to live a safe and secure, interesting and fulfilling life is pretty easy. The scope for improvement across all sectors from systems of governance to systems of food or energy production is almost infinite. This book seeks to develop how these various aspects of our global system might look as we evolve out of 'The Fossil Fuel Age' and into 'The Solar Age'. Later in this chapter, and in subsequent chapters, we delve into my proposed Global Green New Deal and how system change might be delivered in a world radically changed from the present. But first we need to investigate why we do not already have a system that provides everyone on Earth with peace and justice, basic legal rights and freedoms, economic security and the confidence that the ecological and climate catastrophe has been averted.

There are powerful forces deeply interwoven into the very fabric of our existing systems that are very effective barriers to creating the change we need to achieve. These forces opposing us are many and complex. At the most basic level they are the vested interests of individuals and groups who perceive themselves as threatened by the kinds of changes advocated here. These vested interest groups are themselves a complex web of often competing groups. This includes some conservative religious groups, be they Christian, Muslim, Jewish or Hindu. It also includes many of the workers in, and owners of, established industries who see their own livelihoods threatened. It also includes nationalists in many countries who perceive the world in terms of aggressive and threatening rival countries. Probably the most important grouping is the existing rich and powerful people and corporations who see any talk of redistribution and social justice as a threat to their established status. All these groups have a history of intervening to prevent justice or change to occur. However it is the very wealthy and politically powerful who are now the biggest single blockage to progress, especially in the UK and USA. Most of the media is in their hands, so too industry, advertising and public relations, and their lobbying has effectively bought up the democratic process in many counties. Through their control and influence they shape the Overton Window. What they do not approve of is seldom discussed, except on the fringes, among academics, scientists and activists, and in the pages of books, blogs and non-mainstream journals. Almost all television and radio programmes and nearly all the big circulation newspapers are firmly in the hands of those seeking to stop the kinds of changes advocated in this book.

There is another very large group which represents a barrier to change. This comprises people who might be described as the apathetic or disengaged masses. They tend to be focused very much on their own personal lives and show very little interest and empathy with people beyond their immediate circle. Some of these people are caught up in a struggle to keep their heads above water economically and are juggling multiple jobs and childcare commitments. Others who are doing better financially show an 'I'm all right Jack' mentality, and appear simply not to care. To a very large extent I see these people as the product of our current economic system and the media that promotes the values embedded within it. I hope to show in this book how these people might change as they begin to benefit from changes to economics, working relationships and our mass media. This group may be large, and in a democracy they have a lot of power in terms of who they vote for, but they are not in general leading the resistance to change. They tend to be the followers of historical change rather than the initiators of change. They are often driven by fear and they are often easily led by populist politicians.

Interconnectedness & Change
I want to stress the interconnectedness and mutability of things. Politics, economics, societies, cultures, values, beliefs, technologies and individual psychologies are usually studied as separate disciplines,

yet all of course are inter-related and interdependent. Changes to any one create change in all the others and they also all have consequences in terms of the climate and ecological crisis. When climate strikers demand system change they are of course demanding changes to all of these things. In some ways they personally strive to embody the changes they want to see. They are all non-violent and peaceful, mainly internationalist in outlook and into minimizing their negative impact on the world by, for example, not flying, going vegan or refraining from buying too much stuff. There is a saying 'Be the change you want to see in the world'. Processes of change flow both from the individual up to the societal level and also back down again in a complex and ever evolving dynamic. This can all sound very abstract. Let me start with a personal reflection of my family's changing relationships with German people, and the various ways the German character has manifested over time.

In the eighteenth century Germany was made up of numerous small states and the national character was usually portrayed as romantic, musical and dreamy. From the battle of Waterloo and the emergence of Prussia as a military force efficient militarism became a dominant image, enhanced during Bismarck's forceful policies of 'blood and iron'. Rival empire building led to the First World War. My grandfather, still in his teens, fought and was wounded in the trenches. The Treaty of Versailles in 1919 made inter-war Germany virtually ungovernable, and sowed the seeds for the rise of the Nazis. My father was captured on the retreat towards Dunkirk and spent the following five years in a German Prisoner of War camp. Following the end of the Second World War the allies had learnt the lessons of Versailles and under the Marshall Plan Germany was helped to rebuild, a new constitution was established in 1949 and from 1950 the European Coal and Steel Community was established that has since evolved into the European Union.

By the time I lived in Germany during 1977 and 1978 Germany was a very different place. Changes to international agreements, to the German constitution, to politics and to society had fundamentally changed German people's behaviour. From the chaos of the Weimar period, fear and anger had been fuelled by Nazi propaganda to such an extent that people willingly became brutal workers in the concentration camps. By the time I lived there the national character had changed once again. Peace, security, European cooperation, good quality media, a ban on hate speech and three decades of reflection and learning had effectively made the Germans into some of the most ethically sound people in the World.

Changes to constitutions, laws and governmental systems can and do change individual lives and national character traits in dramatic ways. The idea that people are inherently greedy and selfish, or generous and altruistic, is a bit simplistic. We can be, and are, all these things. Human beings are social beings, embedded in ever evolving cultures. Individually and collectively we are highly mutable. As economic and political systems change, so does culture and behaviour. System change is possible, indeed inevitable, and it is up to us to help shape those changes in the most positive direction we can. Peace, justice and human solidarity are fundamental aspects of combating the climate and ecological emergency, and they are achievable, but not under the current dominant political system.

That system is designed to encourage greed, selfishness and the atomisation of society. It rewards short term gains over long term safety, security and sustainability. What we need is exactly the reverse: a system where sharing and human solidarity is paramount, and where the long term future is secure, safe and sustainable.

Political Systems: A New Framework
We need a new political model. This chapter seeks to define what that might look like. This is no easy task. Most historical precedent is of systems that lead to brutality and injustice. Modern turbo-charged capitalism is destroying the planet and impoverishing the many while enriching the few. Socialism, at least as practised under the communist regimes of the twentieth century did its share of ecological destruction while brutally controlling its people. Political discourse has been dominated by this history of left versus right, of capitalism versus

communism. It seems an outdated debate and utterly futile in relation to the climate and ecological emergency.

Instead I want to propose a new political analysis which looks at global politics as currently divided into three rival systems of power and ideology: Pathocratic Populism, the old Establishment Mainstream that dominated much of the late twentieth and early twenty-first centuries, and a newly emerging set of ideas that I'll term the Movement for People and Planet, for want of a better term.

Pathocratic Populism

Pathocracy is a term coined by Andrew Lobaczewski and developed by Steve Taylor in the journal Psychology Today.[7] It explains why people with personality disorders, especially narcissists and psychopaths, are drawn to positions of power. Hitler and Stalin are obvious examples from the mid twentieth century. Donald Trump, Viktor Orban, Recep Tayyip Erdogan, Jair Bolsonaro, Mohammed bin Salman, Sheikh Mohammed bin Rashid Al Maktoum, Rodrigo Duterte and Boris Johnson, Narendra Modi and, perhaps most obviously, Vladimir Putin: all could be argued to fit into this category.

Pathocratic populists are adept at using the prospect of power and glory to reinforce national and sometimes religious pride. Many people are drawn to this kind of leader because they allow them to revert to children, to see the political leader as the all-knowing parent who will lead them through life. In times of confusion, stress and economic hardship people are often drawn to such leaders out of desperation. Such leaders always exert an ever greater control of the media and utilize propaganda to get their message across, with scant regard for truth or reality. They hate real democracy and tend toward ever greater centralization of power and to totalitarian methods of imposing their policies. The old saying that 'power corrupts and absolute power corrupts absolutely' certainly applies to them. Other people lacking empathy, driven by ambition and often also with personality disorders are brought into their governance systems, whereas caring and empathetic people are rejected or leave in despair, so the system becomes entrenched.

Getting rid of leaders such as these is often very difficult as elections tend to be rigged, cancelled or simply non-existent. Sometimes they die in office or are deposed in popular uprisings or military coups, and sometimes normality is restored through the electoral process, as is happening in the USA as President Joe Biden seeks to undo the damage wrought during the Trump presidency.

Very rich individuals and corporations often support such leaders as they can rely on massive favours in return. Where the prospect of making huge profits aligns the interests of business with the interests of pathocratic leaders they will not care what human suffering and ecological damage they cause. The philosopher Alain de Botton defined a demagogue as 'one who protects the rich by getting the poor to blame the weak'. We see this played out daily on our news as Trump blames Mexican migrants for all manner of American problems, while Farage and Johnson use the same rhetoric attacking the migrants who cross the English Channel in tiny rubber boats. The billionaires who own our media love such leaders, as long as they are the leaders of 'our side'.

Most of the pathocratic populists are on the right of politics, but by no means all. Stalin had many of the same characteristics. Hugo Chavez in Venezuela, Ceausescu in Romania, Mobutu in Zaire, Muammar Gaddafi in Libya and Kim Il Sung, Kim Jong Il and Kim Jong-un in North Korea would all in various ways fit into this model of pathocratic populism. They were certainly all despotic. Some, but not all, embezzled vast fortunes and built themselves palaces reminiscent of Versailles. As I write in February 2021 crowds are on the streets of many Russian cities seeking to overthrow Putin, triggered by his latest palace building, his repression of all opposition and especially his arrest of Alexei Navalny. They are also protesting about the economic and political stagnation of Russia and the failure of Putin to act on the climate and ecological emergency. Russia under Putin has become a hollowed out economy, massively geared toward armaments and fossil fuel extraction. Although Russia is in many ways weak, Putin is perhaps the most influential politician in the world today. He is a master of manipulation. Brexit was largely his creation, as the funding streams revealed by Carole Cadwalladr and others show. Boris Johnson is, in part, his puppet clown, whose buffoonery masks the rapid decline of British democracy and the development of a kleptocratic state.

Pathocratic leaders tend to seek power and glory while abysmally failing to deliver the basic economic and ecological security that people crave. Tragically many of the biggest countries on Earth are currently governed by this kind of leader.

The Old Establishment Mainstream

Fighting back the tide of pathocratic populism are the forces of the old establishment mainstream that has dominated global politics for most of the past seventy-five years. It is the politics of Angela Merkel, Emmanuel Macron, Tony Blair, Barrack Obama, Joe Biden, Justin Trudeau and countless others. They represent the forces trying to reconcile a business-as-usual approach to politics and economics while also trying to address the climate and ecological emergency and the rising tide of human suffering. They try to reconcile taking a scientific approach to problem solving with a quasi religious belief in the need for economic growth. They are forever trying to balance irreconcilable forces such as the interests of the fossil fuel industry and the demands of Extinction Rebellion. They are also trying to balance the greed of the very rich and powerful with the needs of the poor and vulnerable. With both the ecological crisis and the social crisis, they are generally prevented from meaningful action by an attachment to the status quo.

This old establishment mainstream has of course its various forms. At one end of a spectrum one might place Margaret Thatcher with her belief in free market economics and centralized political control and at the other the Nordic Model of a highly egalitarian and decentralized society. Thatcher was verging on the pathocratic populist model, while someone like Finland's Prime minister Sanna Marin is about as close to the new emerging movement for people and planet as we yet have leading any national government, anywhere.

Labour, Conservative, Liberal or Social Democratic Parties and most of the political parties across Europe, North America and many other countries traditionally fitted into this broad category. The Republican Party in the USA used to fit in, but since the ascendancy of Trump they have stepped a long way into the world of the pathocratic populists. The Conservative Party in the UK under Boris Johnson is following Trump's lead out of the zone of the political mainstream and into the wild world of pathocratic populism. In the main the EU is still a bastion of the values of this old establishment mainstream, but even here we see the government of Viktor Orban in Hungary edging toward the pathocratic model, and right-wing populists are an ever present threat in many EU countries.

Whereas governance by populists nearly always features the 'strong man' individual leadership model, the more normal mainstream model tends to pay greater priority to competence, cooperation and being good team players. The populists tend to bask in flouting the rules, even their own rules, whereas following established rules and procedures is more highly rated by the mainstream. Corruption tends to flourish more under the populists, although far from rare in the old mainstream governments.

These mainstream governments seem to be under attack as they try to reconcile the irreconcilable forces of the climate and ecological emergency with their desire to protect the very industries and economic interests that are creating that emergency. A new movement is emerging to challenge both the old establishment mainstream and the pathocratic populists.

The Movement for People & Planet

This is a movement largely outside the mainstream of organised political parties. It has no codified manifesto or structure but is made up of a huge array of individuals and organisations. It has some allies within some political parties in many countries but in few governments. It has very clear priorities. The interests of a healthy biosphere are paramount. We have to address the climate and ecological emergency with all the resources at our disposal. The welfare of ordinary people, and especially the poor and marginalized, should be prioritized over the interests of the mega rich. The interests of future generations need to be properly protected. Global politics and economics need to be redesigned to fit these social and ecological priorities.

The School Strikes Movement and Extinction Rebellion are at the forefront of this movement, but there are all those many organisations across almost every country on Earth that Paul Hawken tried to catalogue and many of them would see themselves a part of this movement. This movement is only at a very embryonic stage within organised political parties, but the Sunrise Movement and some of the Democrats they support in the USA such as Alexandria Ocasio-Cortez and Ilhan Omar are very much a part of it. Green Parties across Europe would also see themselves at least in part as the party political wing of this wider movement. Within many other political parties there are some individuals sympathetic to the goals of this movement, although they are usually a minority within their own parties, and there are some governments and countries moving towards accepting the realities of the climate and ecological emergency and therefore becoming more closely aligned with this movement.

Many indigenous tribal communities and peasant communities have sought to resist the forces of colonialism and capitalism for centuries. They have tried to resist land grabs and to point out the sacredness of the land, rivers and the natural world, and that such things simply should not be commoditised. We see in many protest movements around the world now the joining together of these indigenous communities with the global networks of activists for climate and ecological justice. It is increasingly being seen that racial justice and economic justice are absolutely fundamental to climate justice. In many of the world's cities it is the poorest, and often also people of colour, who are suffering most from air pollution and from extreme weather events. It is the very poorest people in places like the African Sahel who have contributed least to climate change and yet are suffering most. Very rich people, who are disproportionately white, and have the highest individual carbon footprints, can insulate themselves from many aspects of the climate breaking down. There is undoubtedly a strong racial justice issue embedded in the climate crisis.

On Twitter I follow hundreds of climate activists from nearly every country on Earth. Many are barely teenagers yet they show a degree of understanding of the science and the need for global grassroots solidarity that I personally find inspiring. They are tremendous networkers. They are part of a vast global network of people and organisations that has no single leader, no single agreed name, no formal organisation and virtually no money behind it, but which seems to me to be the most powerful emerging political force on the planet. It is this that I call 'the Movement for People and Planet'.

Creating Change

One simple test to see where any given politician fits into these categories is to look and see where they stand on economic contraction. It is clear that the fossil fuel industries, and some other industries, need to contract rapidly and politicians sympathetic to the Movement for People and Planet know and understand this. The old established mainstream tends to acknowledge that things need to change as we decarbonise but cannot countenance either outright bans or taxing things out of existence. Their desire to protect existing jobs and industries prevents them from taking the necessary action on the climate and ecological emergency. The populists tend to deny there is a problem, or simply think they can bluster their way though. Only the emergent movement for people and planet sees the need to close many industries and for millions of people to lose their existing jobs. For this to be done in a socially just manner not only do new jobs need to be created but strong social provision is needed to protect the vulnerable, and that requires large scale global redistribution of wealth and resources. There is a fundamental choice: do we prioritise the right of companies to make profits while polluting the planet, or do we prioritise the health of the planet and the welfare of the poor. We cannot do both.

We can of course have prosperity without pollution. That belief is at the heart of this book. The basis for that prosperity will need to be founded upon very different political and economic structures than those which currently exist as the dominant global system. However there are many countries and organizations that are gradually moving in the right direction and offer some degree of hope. We will look at some of these examples of better leadership later, but first we need to explore the scale of the challenge and lessons about the nature of change.

Decarbonising the global economy through a rapid transition from fossil fuels to renewables, from an economy based upon waste and excess to one based upon a modest and circular re-use of resources, and managing our land and soil better: these all look like relatively easily achievable goals compared to the political and economic changes required. However, change can come suddenly, and unexpectedly. The fall of the Berlin Wall and the end of Communism come to mind. The politicians of Eastern Europe knew that their adherence to obsolete Marxist orthodoxy was holding their economies back and that the people had lost any respect for the political leaders. When the Wall came down I hoped the world would unite in a more constructive endeavour to find common solutions to problems. Instead we had the stupid triumphalism of the West saying that they had won the Cold War, as if the contradictions in their system were not glaringly evident even back then in 1989. So we have wasted the last thirty years. Now it is time for the regimes in the West to come tumbling down: they too have lost the respect of the people and are absolutely unprepared to face the macro-level challenges of the climate and ecological emergency. What if they just all collapsed and handed power to The Movement for People and Planet, or if the Movement grew and gained control through normal political channels? In either of these highly unlikely scenarios, what would we do? Are there countries or political models we could learn from? In short, yes there are systems of government to learn from, and things we would want to do to achieve the social and ecological goals that this movement stands for, which this book explores in some detail. However, first I want to point out a cautionary tale from Russian history.

From 1450 and Ivan the Great to 2021 and Putin, Russia has gone through some remarkable changes but has also exhibited some remarkable continuities. From the evolution of Tsarist regimes and the formation of the Russian State, through revolution and communism to Putin and the post-communist state, most political commentators would focus on the enormous changes. But if we look a little deeper some characteristics remain fairly constant. There has been a continuous push for territorial expansion into central Asia and Eastern Europe. Russia has remained a highly centralized and authoritarian state. Human rights and personal freedom have always been restricted and any opposition brutally crushed.

The lesson from all this for us in the Movement for People and Planet must be that we should always remember that changes in a government or a system of government is not the main goal. We need to change something much deeper. It seems to me that kindness and compassion for people and planet are more likely in a system of empowered citizens and communities, structured together into networks of cooperation. No one individual or country should have too much power, money or influence.

Another lesson is that to create a non-violent and more humane world the methods must be non-violent. The history of violent revolutions is nearly always that they bring in violent and brutal regimes. However there are many examples of better systems of governance emerging in post-conflict situations. The critical difference is in the motivation of the people involved. Those who seek to personally replace an existing government and are prepared to use violence to that end tend to be exactly those pathological individuals who are best kept away from positions of power over other people. Whereas those seeking to rebuild in post-conflict situations are usually appalled by violence and seek to cooperate across divisions and to build social solidarity: exactly the kind of empathetic people we do want to be in leadership roles.

Leadership: States & Size

The many places that can in diverse ways be used as models tend to be smaller countries, or regions within larger countries. Size does seem to matter. The ten biggest countries by population are China, India, the USA, Indonesia, Pakistan, Nigeria, Brazil, Bangladesh, Russia and Mexico and none of these provides models of leadership I would want to follow. Leopold Kohr in 'The Breakdown of Nations' argued that large countries tend to be more aggressive, less happy and with a population feeling alienated from its government. Smaller units work better, where the leaders are in closer touch with their own people and are less likely to be consumed by hubris. Not that being small is any guarantee of good governance, but being too big does tend toward the kinds of imperialistic behaviour that are detrimental to the kind of progress humanity requires.

I became intensely aware of this in the 1960s. The American bombing campaign in Vietnam and the brutal Russian suppression of the Prague Spring were two of the formative experiences of my youth. Both followed a long line of extreme aggression by the two superpowers. The USA has long trumpeted democracy while covertly backing coups and installing far-right regimes abroad, often with the backing of the UK and other governments: Iran in 1953, Guatemala in 1954, Congo in 1961, Brazil and British Guiana in 1964, Ghana in 1966, Indonesia in 1967 and Chile in 1973, to name but a few. The Soviet Union had a similar pattern, brutally suppressing attempts to build greater freedom and grassroots democracy in East Germany in 1953, Hungary in 1956 and then in Czechoslovakia in 1968. The Chinese annexation of Tibet, the Indian occupation of Kashmir and its wars with Pakistan, the Indonesian suppression of the people of West Papua: the pattern of big countries using bombs and tanks to attack largely unarmed civilians was a pattern all too familiar. It continues today with Russian involvement in Ukraine, Crimea, Belarus, Kazakhstan and Syria. Saudi aggression in Yemen has been made possible with weapons supplied by the USA and the UK.

There were of course individuals offering a more peaceful world view. Gandhi was, and the Dalai Lama remains, inspirational in this regard. Peace movements emerged, with the CND in UK and the anti-war movement in the USA, but they did little to change the behaviour of their governments. Trying to find governments actually committed to peace is revealing. The Institute for Economics and Peace publishes an annual league table. In 2020 the most peaceful place was judged to be Iceland, followed by New Zealand, Portugal, Austria and Denmark.[8] The Institute for Economics and Peace tables seem to me to be heavily skewed towards wealthy countries. These top five are all small, stable and relatively wealthy countries. I am equally impressed by a diverse range of other small countries.

Costa Rica emerged from a civil war in 1948, abolishing its army, and investing heavily in health and education. What it has achieved over these last 73 years has been very impressive. It is now one of the most peaceful and safe countries in Latin America, with excellent health, education and social provision. It is also impressive in that it generates over 99% of its electricity from renewables, including hydro, geothermal, wind and solar, and it has been more successful in reversing the destruction of its rainforest ecosystems than probably any other country on Earth.

The improvement in governance in Uruguay has also been impressive. Emerging from military juntas in the 1980s it has seen extraordinary progress across many areas. It has cut its defence spending and like Costa Rica invested in health, education and social provision. Jose Mujica was one of the most impressive political leaders of any country over these last decades. What he and his Broad Front political grouping achieved is quite remarkable across many fields, and personally his humble and generous lifestyle provides a role model for good leadership. He gave away 90% of his salary, insisting he did not want to earn more than the average Uruguayan worker.

Bhutan is another interesting country. In 1972 their king declared 'Gross National Happiness is more important than Gross Domestic Product.' For nearly fifty years it has been developing a 'Gross National Happiness Index'. It has been one of the pioneers of new economic models. There are dozens of other individual small countries that have achieved impressive results in a single area or across the broad range of their development.

One could of course argue that small countries will be less aggressive than large countries as they simply do not have the power and capability to invade and dominate other countries. These small and peaceful countries I have just mentioned do seem to offer a way forward which the big countries that I discussed in terms of their hubris and aggression do not. Maybe it is time to shift power away from the model of the big nation state with their traditions of imperial ambition and towards networks of smaller units linked together into new global patterns of cooperation?

Leadership through Networks

Networks of co-operation are in many ways the polar opposite of the strong man leadership model so beloved by the pathocratic populists. There are a growing number of such networks striving to move toward better systems of governance. None of them is perfect, or as good as they need to become, but they do provide a model upon which to build. We will examine a few of these existing networks and then I will introduce an idea of a global network of small decentralized communities linked together in a global network of co-operation through which my proposed Global Green New Deal would be delivered. So, let's start by looking at some of those existing, imperfect but also inspirational, organisations.

The European Union is certainly imperfect. It is very much a product of the old establishment mainstream, trying to balance all manner of conflicting goals, but what it has achieved is still remarkable. Out of the carnage and destruction of the Second World War peace and prosperity has been built. The 75 years from 1946 to 2021 have been the most peaceful in European history. The previous 75 years included the Franco-Prussian War, and the First and Second World Wars. The importance of that turnaround cannot be overstated. Achieving that peace and prosperity was based on peaceful co-operation and respect for diversity while reinforcing solidarity. These are vital lessons to now take forward on to a global stage.

There have been many initiatives and networks linking individual towns and cities seeking to bring about a change toward a more ecologically sustainable and socially just future. Notable was the Rio

Earth Summit held in Rio de Janeiro in 1992, out of which evolved many Local Agenda 21 initiatives. In 1994 the Aalborg Charter was signed in Aalborg, Denmark, which has evolved through the Aalborg Commitments and into the Sustainable Cities Platform. Thousands of local authorities from dozens of countries have participated in this process over many years. This process provides a model of networked, bottom-up change. In countries where political and economic power is highly devolved the possibilities to create change at this local level are immense. In highly centralized counties, such as the UK, it is hard for people to understand just how much can be achieved by this kind of empowered local leadership.

A decade or more ago I visited the small Austrian town of Güssing, which had achieved great things, creating jobs and new industries while reducing its carbon footprint and improving the local biodiversity. Güssing, with a population of just under 4,000 people, is about the same size as some of the market towns of Herefordshire, such as Bromyard, with a similar population but a budget only about one fiftieth as big as Güssing's. Our parish councils cannot do a great deal: they do not have the money to employ people and get anything much done. In Edwardian Britain our local councils had the power to get things done in a way that for decades now they have not, due to continual budget cuts. Empowered local communities are at the heart of my Global Green New Deal, and for countries such as the UK that must mean a vast shift of resources from central to local government.

The Nordic five: Norway, Sweden, Finland, Denmark and Iceland, are probably the best governed countries on Earth, and effectively cooperate together and learn from each other. They are by no means perfect, but they have achieved a lot. In the 1860s both Sweden and Finland suffered terrible famines with mass emigration. During the nineteenth century they were poor and unequal; now they are all prosperous and more equal than any other region on Earth, having largely conquered the problems of poverty and inequality. They are beginning to make real strides toward becoming more ecologically sustainable, but still have a long way to go on this front. The improvements in living conditions and in wealth distribution were slow and incremental. The era when the climate and ecological emergency could be solved by slow and incremental change is over, and now change needs to be much more rapid. However we have an enormous amount to learn from the Nordic region. Anu Partanen's book 'The Nordic Theory of Everything'[9] details how society is organised across the Nordic Region and contrasts this with how things work in the USA. We have a lot to learn, and her book is one of the key inspirations for my ideas about the detail of how my Global Green New Deal might be organized. So, for example, later in this book we will investigate setting up a global health service, and global educational provision, inspired by what has been achieved in Finland, and adapted to local needs and to the new opportunities as we transition out of 'The Fossil Fuel Age' into 'The Solar Age'.

The Wellbeing Economy Alliance[10] is attempting to build a movement that seeks to prioritize wellbeing and to act on the climate and ecological crisis. It includes individual members, many organisations and some governments. It is still quite embryonic but seems to have great potential, in some of the same ways that the Aalborg Process has. Part of the Wellbeing Economy Alliance is Wellbeing Economy Governments,[11] which so far includes the governments of Scotland, New Zealand, Iceland, Wales and Finland. Five very interesting small governments, including two of the constituent nations of the UK, here plotting a developmental course diametrically opposed to all that the UK government under Boris Johnson is doing. Post Brexit UK seems to be on divergent paths and may well lead to its breakup. Boris Johnson, following the pathocratic populist model, is trying to push a nostalgic imperialist view of the world and a revival of English dominated British nationalism. Meanwhile Nicola Sturgeon in Scotland and Mark Drakeford in Wales are inching slowly away from the establishment mainstream positions towards more ecologically literate and socially just egalitarian models stressing the importance of networked collective leadership.

Since 2014 the High Ambition Coalition[12] has been leading the way through the COP climate meetings in Paris, Madrid and elsewhere.

Interestingly leadership has come from several small vulnerable island states such as The Marshall Islands, Saint Lucia, Fiji and Jamaica, and countries such as Costa Rica and New Zealand, several in Africa and most of the European Union. Mia Mottley of Barbados was one of the outstanding figures of the Glasgow COP in 2021. Tragically the whole COP process has been derailed by the intransigence of the usual big players, such as India, China, Brazil, Russia and Saudi Arabia. The UK's role under Boris Johnson was as chaotic and full of empty bluster as one would expect. Within the context of leadership I am discussing it is important here to note that the key leaders were small and vulnerable nations coming together in networks.

Time, Deliberation & Democracy

Time is not on our side. Most of the examples listed above have achieved impressive but still inadequate results over long time spans. The multiple crises we now face demand rapid and radical change.

We have to be clear how much is stacked against us. The scale of the changes required and the time available looks close to impossible. Most of the media, most of the politicians and much of the population are wedded to ideas, policies and the incumbent industries and technologies that are driving us toward destruction. The fact that many councils and parliament have declared climate and ecological emergencies and yet recently councillors of all three mainstream political parties in the UK voted in favour of a new coal mine in Cumbria says it all. None of them seems to understand the situation we are in. What can we do?

Extinction Rebellion is putting much faith in Citizens' Assemblies. These are groups of ordinary people representing a cross section of the population who listen to expert witnesses and take time to consider complex arguments, and then, like a jury in a court case, make a collective decision on the best way forward. Some governments are beginning to initiate Citizens' Assemblies, but usually with no power to implement anything, simply as bodies to make recommendations which governments are then free to ignore. Extinction Rebellion is calling for Citizens' Assemblies to be given real power. They certainly will have some kind of a role, but don't seem to me to be a complete substitute for democratically elected governments. Although many governments have been so corrupted by big money and fossil fuel lobbyists, not all have. It is probable that Citizens' Assemblies will be developed into effective tools for reinvigorating democracy in those places that already have very good decentralized democratic structures, exactly those same places that are already participating in things like the Aalborg Process and the Sustainable Cities Platform. Someone like Boris Johnson will set them up to look good, but probably totally ignore their findings.

In the UK and USA we are at a particular disadvantage, hampered by ridiculously undemocratic political systems using first past the post elections. This promotes just two rival parties and thus excludes smaller, newer parties, which often are exactly the people bringing new ideas into politics. The two rival parties become polarized and utterly unable to cooperate effectively. In countries with proportional systems there tend to be many more parties in parliament, usually a couple of dozen or so, with new parties constantly forming as new ideas emerge. Coalitions tend to be the normal system of forming governments. This encourages much more co-operative and reasonable dialogue and policy formation.

Big money has had a particularly undemocratic influence on American politics, and at both the federal and state levels politics has become polarized on party lines, and all Republicans and most Democrats only really represent the views of their corporate sponsors. Interestingly, at least in some places, the city and town council level politics in the USA seems very much healthier with active cross party or non-party co-operation and community participation.

In the UK our system is hopelessly outdated. One short term goal must be to establish some system of proportional representation. I would urge readers to support Make Votes Matter which is a great little organisation working to build support across the political divide.

Because the extent of the multiple crises and the time available to avert catastrophe, much of the pressure for change will have to be on the streets and in other processes outside the traditional democratic process. For that reason supporting groups like Extinction Rebellion, the School Strikes movement and countless others groups is vital. However it is important not to ignore the formal party political processes. In the May 2019 Herefordshire Council elections[13] we managed to defeat a long established Conservative Council with a coalition of Independents, Greens and a small local party called 'It's Our County'. The effect has been dramatic, with, for example, the cancellation of a major road building scheme and a much greater emphasis on walking, cycling, public transport and the desire to establish car sharing clubs. It is important to engage with party politics, however imperfect the system undoubtedly is.

One critical paradox of our situation is that of time. The climate and ecological crisis demands an urgency of action, but the actions must be guided by a kind of very long term thinking that is outside the scope of our current system. Most politicians are concerned with the next day's headlines and with the next election, and business is focused on quarterly and annual profits. It is of critical importance for us to act now in the interests of future generations. Roman Krznaric[14] points out how our present decisions have dumped problems on to future generations in a way that is undemocratic. The unborn have no vote or voice. A global movement is growing against this tyranny of the now. Our Children's Trust has taken the USA government to court asserting the legal rights of young people and future generations to a safe climate and a healthy atmosphere. This has sparked a wave of similar court actions in dozens of countries. It was a tradition of many Native American communities to think in terms of the seventh generation, and this long term consideration of the needs of future generations is at the heart of the movement for people and planet. We have to 'Act Now' in the interests of these as yet unborn people.

A plan for radical decentralization and global solidarity

There are many examples of better governance at the local level, as typified by those participating in the Aalborg Process. Aalborg itself is a small city of 217,000 people. Herefordshire where I now live has a population of 191,000; the London Borough of Bromley where I grew up, a population of 331,000. If we think of a typical local authority as having an average population of 200,000 while there are 7.8 billion of us humans, that would suggest, at least theoretically, some 39,000 local authorities averaging about this size. In some sparsely populated regions these communities would be spread over huge areas, whereas in the megacities many such communities would be linked together within a larger city authority, as is currently the case with the London boroughs or the Parisian arrondissements.

If we imagine these 39,000 local authorities having vastly increased resources channelled through them, and powerfully networked together to learn from each other in the kinds of ways promoted in the Aalborg Process, a vision of a radically decentralized networked structure begins to emerge. It is with this kind of empowered local democracy that I envisage my proposed Global Trust for People and Planet working to develop projects capable of delivering the multiple objectives I have outlined in my proposed Global Green New Deal. Each of these local authorities would use tools such as Citizens' Assemblies to involve their communities and to draw upon the collective wisdom potentially available within their communities. They would also be constantly learning from other localities and they would have the resources to draw on the knowledge and expertise that exist in many parts of the world. The Global Trust for People and Planet could have as one of its key tasks the dissemination of best practice.

I see this radical decentralization going hand in hand with invigorated global cooperation.

Back in 2003 George Monbiot wrote 'Age of Consent: A Manifesto for a New World Order' in which he argues that humanity may be on the verge of a metaphysical mutation, shifting our allegiance from tribe or nation to species. I would argue that the climate and ecological emergency forces us to act for the collective good of our entire species, or our long term viability as a species is seriously at risk. Monbiot argues for a directly elected World Parliament, for the UN and other global agencies to be transformed into much more democratic and egalitarian structures. It is a book full of good ideas, many of which could be implemented as part of the system change that is at the core of this book.

In 'The Nordic Theory of Everything' Anu Partanen combines personal anecdote with rigorous academic research to compare and contrast Nordic and American systems of governance and how their societies are structured. She is a Finnish woman living in New York. The contrast between Finland, or more generally the Nordic Region, and the USA shows how strange and dysfunctional much of the American system is. She seldom uses the terms socialism or capitalism, preferring to focus on the detail of policy and look at the intention and purpose behind it. She draws on Trägårdh & Berggren's 'Swedish Theory of Love' and proposes a 'Nordic Theory of Love', showing how the wellbeing and healthy development of children and of society is at the heart of Nordic policy making. It explains why so many aspects of Nordic society seem to function so well.

In my Global Green New Deal I think we have much to learn from this Nordic model, and perhaps Anu Partanen's concept of the Nordic Theory of Love could be further extrapolated into a Global Theory of Love, providing Nordic quality health care, education, social provision and governance for a global population. The proposals I make in my Global Green New Deal would channel vast sums of money away from multinational corporations and extremely rich individuals, and also, but perhaps to a lesser extent, away from national governments. Huge levels of investment would be made in transforming the lives of poorer people and in reinvigorating local governments and global institutions, and Monbiot's and Partanen's books provide inspiration.

We will look at how the Global Trust for People and Planet could be used to help establish projects in all our global network of 39,000 local communities. The funding for these projects would be massive, amounting to many trillions of pounds, dollars or Euros per year. It is through this network of projects that massive improvements in health and education and in human wellbeing could be achieved. Each project would seek to be carbon negative and biodiversity enhancing. They would help drive forward the rapid transition from 'The Fossil Fuel Age' to 'The Solar Age'.

Part Two: Society, Economics & Work
Population
Many people worry about the sheer size of the global population. There are currently nearly 7.8 billion of us, and this figure is projected to grow to 10 billion by the late 2050s.[15] The collective demand for resources and the ecological impact of our species is calculated based on the individual human demand and impact multiplied by the number of people. Population concerns from Malthus onwards have focused on the idea that there was a relatively fixed carrying capacity of the world. Malthus expected increased rates of famine as the population began to exceed one billion. We are now at 7.8 billion people. We still have food shortages, but not caused by lack of supply but by inequality and waste. We produce more than enough to feed everyone, if we just distributed it differently. We could feed a lot more people and do it with very much less ecological damage, but to do so diets and systems of production and distribution would need to change, but they need to change anyway, and the changes are quite doable. Food supply ought not to be a limiting factor in human population growth. Other concerns about population focused on finite resources like coal and oil, and still other concerns around climate change and ecological impact.

I want us now to reconsider the situation, and imagine if the net ecological impact of each individual was beneficial to the planet, then it could be argued that the more humans there are the better. I would not go quite so far as to argue that a higher population would make things better, rather that if we can transform our systems of agriculture and industry, of economics and politics, which we must do anyway, then the size of the total global population would not necessarily be a problem.

If we look at carbon emissions, a typical billionaire might be emitting a thousand tonnes, or more, of carbon dioxide each year and an African subsistence farmer might, on average be emitting one kilo or less.[16] One person emitting at least one million times as much as another clearly is not fair. In our current situation of multiple crises it is not food and energy that are putting limits on the carrying capacity of the Earth, but rather the ability of the planet to deal with the pollution that we generate. Highly polluting lifestyles must come to an abrupt end. The existence of anyone anywhere living a billionaire lifestyle is a threat to all people everywhere.

It is possible to imagine every human being on Earth having an individual carbon footprint of below zero. In this envisaged scenario the individuals would also be making minimal demands on finite resources such as metals and water due to the change from a linear to a circular economy, and humanity would no longer be a threat to precious ecological habitats, but rather their custodians. Such a society only becomes conceivable with very different systems of economics and politics. It would inevitably be very much more equal.

In the last epochal shift as the industrial revolution ushered in 'The Fossil Fuel Age' not only did global population rapidly increase, it also moved. People have always moved to where resources are most abundant and the greatest economic opportunities exist. The industrial revolution of the eighteenth and nineteenth centuries saw people moving in large numbers to the emerging cities of the newly important coalfields. As solar power is rapidly displacing fossil fuels already, and will do so ever more quickly, the hot dry deserts of the world will take on a new significance. They will be where energy and land are cheapest, where some of the most exciting new economic opportunities are emerging, and where new cities may well be built. In subsequent chapters we shall explore how these cities could be provided with food and water, and how such cities could be the crucible of wider changes in society. Just as much of the technical, intellectual, political and social progress associated with the last industrial revolution came from the new coalfield cities, so much of the progress associated with 'The Solar Age' may well emerge from these as yet to be built cities in the desert.

*

Resources and the Human Resource

Throughout the many thousands of years of the Stone Age, flint was a highly valued resource, which now we hardly think of as a resource at all. During 'The Fossil Fuel Age' we have seen the emergence of stock markets and the increasing value placed on deposits of coal, oil and gas. Over recent years the value of many coal, oil and gas companies has collapsed as the global economy has just begun to pivot away from fossil fuels towards renewables. This process will inevitably continue and speed up. We now understand that clean air, fresh water and a well functioning biosphere are absolutely vital, yet still our economy attributes no value to them. Clearly our way of giving value and significance has to change, and change profoundly.

Our economic system places very little value on people. Economic significance has been given to demand, as expressed through ability to pay. So the whims of billionaires take precedence over the survival needs of poor people. Of course this needs to change, and one half of the equation is to tax the mega rich out of existence. The other half of the equation is to invest heavily in ordinary people, and their skills, energy and enthusiasm will be of critical importance in achieving the tasks ahead. Humanity is our most powerful resource in doing all that needs to be done to turn the global economy around.

Historically, countries have lifted themselves out of poverty in large part by investing in their people. Finland and Sweden over the last one hundred and fifty years, or Singapore over the last seventy-five years, have conquered hunger and poverty by this method. They now have excellent systems of health care and education, and populations capable and competent to innovate, plan and execute projects of all kinds. It is high time we did this on a global scale. My proposal for a Global Green New Deal would include global provision of excellent health care and education, and most importantly of all, the creation of worthwhile, meaningful and enjoyable work, doing all the things that need to be done to reduce carbon emissions, restore biodiversity and transform the global economy. We will return to the task ahead and see how redefining work and redeploying the global workforce is of critical importance. First we need to look at the economy: what is it and what is its purpose.

Markets and Masters

'The markets make a good servant but a bad master, and a worse religion.'[17] This quote from Amory Lovins cuts to the heart of what economics is all about. Buying and selling, bartering and trading go back thousands of years. The point of such trade was to exchange what one had in surplus for what was of use, but less readily available to produce oneself. As the mercantile system developed into early forms of capitalism the point of production, trade and markets gradually changed into the creation of profits to be reinvested in ever larger trading ventures. That required the invention of stock markets and ever larger corporations commodifying ever more sectors of the economy, of the planet and of human life. Slavery became a turning point. Profits could be made, but questions arose. Was it ethical, and should it be made illegal? Factory conditions in the early stages of the industrial revolution were dirty, dangerous and the hours were inhumanely long. Factory owners could make profits, but at the cost of the health and lives of their workers. Gradually legislation came in to outlaw slavery and to force higher standards on to reluctant factory owners. Markets could be used to make profits, but they needed to be regulated in order to bring in health, safety and ethical issues.

Underpinning global history for the last couple of centuries has been a battle over what role markets should play. It has generated an array of opposing ideological positions and led to numerous wars. On the one hand there have been those in favour of unrestricted markets, who at their extreme we can call the 'Market Fundamentalists.'[18] At the other end of a spectrum we have those opposed to any form of market, and who wanted the state to control all sectors of economic activity. These people have been described as socialist, or communist, but perhaps most accurately of all we could refer to them as 'Anti-market Fundamentalists'. These two rival economic fundamentalist positions are as inhumane as each other. Stalin's grip on the Soviet empire was brutal in the extreme, as was Mao era China. The market fundamentalism so dominant in the UK and USA over the years from 1980 to 2022 has lead to a similarly dysfunctional and inhumane system. This may sound abstract. If we look at how these rival ideologies have shaped food production and consumption we see how stupid they both are, and how a better solution is immediately apparent.

The Soviet and Chinese anti-market fundamentalists forced land collectivization onto a reluctant peasantry. Decisions about what crops could be grown and where and how they should be grown, processed and distributed were centralized into governmental offices in Moscow and Beijing. The actual people working the land were disempowered, so those in daily contact with the soil conditions and local weather were not the people deciding how to respond to these changing conditions; instead these decisions were largely made by distant officials. Famines usually follow centralized government controlled land collectivization. Life expectancy often falls.

Under the market fundamentalists any notion of morality beyond the market is stripped away. Need is expressed as willingness and ability to pay. As market fundamentalism inevitably leads to ever greater levels of inequality, a class of mega rich becomes ever richer while mass impoverishment ruins the lives of the vast majority. In terms of the food economy, environmental safeguards are cut back and production maximized. But this production is not aimed at feeding all humanity but at catering to the dietary whims of the very rich and generating profits for agribusiness. Extravagantly decadent restaurants and food banks both flourish. Rainforests are cleared to make way for agribusiness to make profits but at devastating cost to the environment. Over-processed junk food is made, advertised and sold making profits for companies but undermining human health. Poverty, malnutrition and obesity have become increasingly widespread, especially within supposedly rich countries like the USA and UK. Life expectancy in the USA is now falling, in part as a direct consequence of these policies.[19]

Producing enough good quality food for everyone and doing so sustainably is actually not difficult in theory, but it is totally impossible under either of these fundamentalist systems. In both the vast majority of ordinary people are disempowered, and power is concentrated into ever fewer hands. In the market fundamentalist case it is the very rich, and in the anti-market fundamentalist case it is the centralized political party machine.

A better economy would utilize markets as a tool, but not make a master or a religion of them. Let us imagine a sufficiently regulated economy that safeguarded the environment and employed redistributive taxation to ensure ever increasing levels of global equality. In such a society everyone would be able to express economic demand in a roughly equal manner. Everybody would be empowered to buy the food they needed, and farmers would grow what this new and egalitarian market deemed to be in demand. Governmental structures, from the global to the local, would set regulations to minimize pollution and any form of environmental damage. Farming policy would fund and support systems that were good for human health, from universally available fresh local organic produce to cookery classes and outdoor adventure activities. Incentives would be put in place to ensure soil based carbon sequestration and ecological renaissance, but individual farms would decide their own planting and harvesting regimes within this regulated and incentivised framework.

During the so-called 'Global Age of Capitalism', from 1945 to the early 1970s there was general agreement among Western governments that markets were useful, but needed to be moderated by legislation to enforce health, safety and environmental standards, and taxation was needed to redistribute wealth to make society fairer. In the late 1970s cracks in the Keynesian consensus appeared. Stagflation, the combination of economic stagnation with high rates of inflation was at the root of the Keynesian crisis. It was the crisis that the market fundamentalists were waiting for and from 1979 we see with Ronald Reagan and Margaret Thatcher a move toward market fundamentalism and growing inequality that has dominated the UK and USA for the last forty years, and has grown to dangerous levels during the Trump-Johnson pathocratic populist era. Just as the communist, or anti-market fundamentalist, system came crashing down in 1989, now is the time for the market fundamentalist systems to crumble and fall.

We need a new system that embodies global social justice and fairness, climatic stability and ecological resurgence, and uses markets as a powerful tool to achieve these objectives. Markets without strong regulatory frameworks have become a bad master. To the market fundamentalists, markets have become like a religion, with results that have become socially and ecologically devastating. New global regulatory frameworks are absolutely necessary if we are to address the multiple crises we now face.

Growth, Degrowth and 'The Doughnut'

With the development of capitalism and the need for ever expanding markets came the notion of economic growth. This growth has made us more prosperous, which in many ways is of course a good thing, but it has come with terrible social and environmental costs. The pursuit of economic growth as a central goal of government policy has really gained overwhelming importance over the decades since the end of the Second World War. Since the early 1970s a growing band of authors have pointed out the absurdity of indefinite growth on a finite planet. 'The Limits to Growth' by Donella Meadows and others came out in 1972 and 'Steady-state Economics' by Herman Daly was published in 1977. Since then there have been hundreds of critiques of economic growth as a central policy objective, but almost all governments have stuck to the dogma that economic growth is absolutely essential.

Economic growth is a very clumsy indicator. It measures how much money is spent within an economy, but does not distinguish between things that people might want, like better schools, from things that nobody wants, like car crashes. Car crashes and building schools both generate jobs and both involve the use of finite materials, but they do not amount to similar increases in wellbeing. Over recent decades there have been many attempts to create ways of measuring wellbeing. Some governments are just beginning to make this a central governmental policy objective, replacing economic growth as the prime indicator of the success of the economy. Bhutan was decades ahead of the field, with the king declaring that 'Gross National Happiness' was his key objective way back in 1972. The Wellbeing Economy Alliance is a global collaboration of organisations and individuals promoting this wellbeing based focus to economic policy. Part of this wider movement are a handful of governments, led by Scotland, New Zealand, Iceland, Wales and Finland, and as was previously mentioned, they are all small countries. It will be very interesting to see how many other governments sign up to this wellbeing focused agenda. That might be one of the clearest possible indicators of positive change in the world.

Kate Raworth's book 'Doughnut Economics' was published in 2018, and has been highly influential. It uses a brilliantly simple graphic shaped like a doughnut. There is a safe, sustainable and just space for humanity to occupy. There is an environmental ceiling beyond which lies the zone of environmentally unsustainable overconsumption. There is also a social foundation below which lies socially unsustainable poverty and suffering. Kate Raworth describes herself as growth agnostic. Her focus is on simultaneously reigning-in overconsumption, which would imply degrowth, and with alleviating poverty and suffering, which would imply growth. Whether this amounts to overall economic growth or contraction is hardly relevant, as growth as an indicator is such an inadequate tool.

I recall talking many years ago with Stephen Harding, the ecologist at Schumacher Collage, about his ideas about economic growth. He divided the economy into four categories: suicidal growth, intelligent growth, suicidal contraction and intelligent contraction. Much of our current growth is suicidal in the sense that it is leading to ecological catastrophe, or to social division and wars. Other aspects of growth are good, which produce real gains in human wellbeing while safeguarding or enhancing the environment. In Harding's terminology it would be suicidal to contract these good areas of the economy. The category that is perhaps most interesting is that of intelligent contraction. For our growth obsessed governments, planning to shrink any aspect of the economy is always difficult as it implies the end of some people's jobs and the end of the chance for some companies to make money. Given the climate crisis and the improving technological capacity and falling costs of renewables it is obvious that all sections of the fossil fuel economy need to contract. That would be the intelligent thing to do, but our governments struggle to implement this as it goes against their belief in growth.

Many economists, like Mark E Thomas, still believe that overall growth can and should still be a policy objective, but like Raworth and Harding, he sees some growth as being good and other growth as being bad. Thomas portrays growth as fitting into three categories: good growth, growth that can be made into good growth by modifying the processes involved, and irredeemably bad growth. The challenge for economists, politicians and society will be to sort out which projects and technologies fit into which of these three categories.

Gradually the world is acknowledging that some technologies are just irredeemably bad. Burning coal without full carbon capture is now in most people's thinking unacceptable. However this does not stop some governments and companies still building such obsolete and polluting power stations. That is why there is a clear need for global regulation and enforcement, and help with incentives to develop better systems to save energy wastage or to generate cleaner electricity.

In thinking about almost any project or technology there is almost always scope to improve it if we take into account the full impact of it on the wellbeing of people and planet. Under a growth dominated political system these considerations are seldom prioritized. If money can be invested and profits made those are sufficient grounds for the project to go ahead.

The tricky part for governments and society is to work out which technologies and projects can be transformed from bad to good growth. Some might argue that by fitting carbon capture technology to a coal fired power station it could be transformed in this way. Others, me included, would argue that although this would make it less bad it certainly would not make it good. Less polluting, cleaner and more efficient technological alternatives exist. For the same level of financial investment a range of energy saving projects could be initiated and renewable energy co-operatives established which would tick a greater number of positive outcomes from a wellbeing perspective.

Triple bottom line accounting seeks to broaden accounting systems to take full account of the social and environmental aspects along with the economic aspects. It is a powerful tool to refocus economic decision making away from a 'growth at all costs' agenda to a more wellbeing focused agenda.

The proposals in my envisaged Global Green New Deal would imply an abandonment of growth as an economic goal. Allocation of funding and project development would use triple bottom line accounting and a very broad focus on planetary and human wellbeing. With these criteria in mind almost any technology and project can be improved. So let us return to the concrete example of global energy supply. If human and planetary wellbeing was the focus we could provide everyone on Earth with more or less the same access to energy usage. All use of fossil fuels would have ended and all 7.8 billion us would be members of a global network of renewable energy co-operatives where the financial benefits were evenly distributed, and the technological components would all be part of a fully circular economy designed for continual reuse. Whether all those changes in total would have amounted to economic growth or contraction is hardly relevant. They would have amounted to a vast improvement in human and planetary wellbeing, and that is of existential importance.

Equality and Inequality

The extremes of wealth and poverty that exist now, and have long existed, struck me as fundamentally wrong even as a child. The more I have studied the social sciences the more this basic gut feeling has been reinforced. The data reveals how dysfunctional this is for any country, and for the world. Richard Wilkinson and Kate Pickett published 'The Spirit Level: Why More Equal Societies Almost Always Do Better' back in 2009. In it they presented the most overwhelming amount of evidence to back up their hypothesis that more equal societies achieve very much better outcomes across a very wide range of indicators.

The simple idea that it is wrong for some people to suffer material poverty and hunger and others to amass excessive wealth and power underpins a vast range of thinking and practice, from early hunter-gatherer bands to the roots of all the world religions, from Marxism to the politics of Mahatma Gandhi, Franklin D Roosevelt, Martin Luther King, Clement Atlee and Nelson Mandela.

In the so-called 'Golden Age of Capitalism' after the Second World War significant progress was made in many countries as tax rates were high enough to make substantial improvements to public services possible, from schools and hospitals to housing and social welfare. Life expectancy and living standards rose. All these positive trends have been undermined or reversed in the era of tax cutting right-wing politicians from the time of Thatcher and Reagan to Trump and Johnson. Now it is time to re-invent big government.

Today the focus needs to be on the governance of the planet and of the whole human family. We can and must reverse the climate and ecological emergencies, and the tools we need to do this can simultaneously be used to alleviate or eradicate poverty and hunger, which in themselves are not difficult goals to achieve given sufficient funding. The ideas I am promoting in this book are a call for big government, but unlike most authors my view is that most of the fund-raising and spending should be at the global and local levels of governmental structures. My envisaged Global Green New Deal would achieve massive scale redistribution and wealth and power.

Inequality is not just an issue of income, wealth and power; it is also an issue of resource use and climate change. Most African subsistence farmers will have carbon footprints of only a few grams, and some might even have tiny net negative emissions if they are building up soil carbon or planting trees. It seems apparent that all, or almost all, billionaires have vast carbon footprints. In a study by two economic anthropologists from Indiana University they looked at the carbon footprints of twenty billionaires and found that they varied from Michael Bloomberg's relatively modest 1,782 tons to the staggering annual emissions of Roman Abramovich at 33,859 tons.[20] A recent report from Oxfam stated that the carbon emissions of the richest 1% are more than double the emissions of the poorest half of humanity. [21]

Humanity currently emits around 40 gigatonnes of carbon dioxide every year. That is 40,000,000,000 tonnes. There are 7.8 billion of us, so that works out at about 5 tonnes per person per year. We need to reduce these 40 gigatonnes to zero as fast as possible. As average emissions vary a lot between countries and even more between individuals, the changes to some countries and to some individuals' lives will have to be more extreme. The typical billionaire lifestyle has to come to an abrupt end.

Climate justice is clearly an issue of racial and economic justice. The people most vulnerable to climate change and extreme weather events are those in low lying areas like Bangladesh, and in areas prone to droughts like the Sahel; exactly the people whose emissions are the least. Within rich countries like the USA and UK it is the poorest people who are disproportionately of Black and Asian origin who are most exposed to air pollution and to climate change related flooding.

My proposed Global Green New Deal would seek to establish equal rights to clean air and water for everybody. Individual carbon rationing would need to be brought in and the most polluting technologies banned or taxed out of existence. A total ban on private jets and on super yachts would be a good place to start, and this of course would almost exclusively affect billionaires. Redistribution must not just be about income and power but equality of resource use and equal rights to breathe clean air. The vast majority of humanity would benefit from such radical redistribution. We will never conquer the injustice of poverty without tackling the unsustainability and unfairness of excess.

Philanthropy or taxation
People have always wanted to help the poor, the sick and the vulnerable, and for a very long time there has been debate over whether philanthropy or taxation is a more effective tool. In small hunter-gatherer bands powerful obligations to share were the norm, but as social groups

got bigger other methods beyond face to face transactions became necessary. Most if not all religions organised tithes to be collected to help the poor, which in a sense were a form of taxation. Ever since states, empires and governments have existed they have used taxation of one sort or another, but sadly this was often more to fight wars than to help the poor. Individual giving or philanthropy goes back just as far as taxation. Some countries and systems promote one more than the other and again Anu Partanen's book 'The Nordic Theory of Everything' is revealing about the extraordinary contrast between the USA and Finland and the Nordic region. The Americans favour individual voluntary acts of philanthropy, the Nordics taxation and legal rights.

Let us look at the issue of homelessness. In the USA there is, and has been for decades, a vast network of charities whose very purpose is to help the homeless, yet the number of rough sleepers appears to be going up. There are over half a million people either sleeping rough or staying in temporary night shelters on any given night. Most of these people are fairly short term homeless, but a substantial minority are long term chronically homeless people. Beyond this there are a vast number of people in overcrowded, insecure and temporary accommodation.

In Finland no such charities are needed. Finland has pretty well sorted the problem. Homelessness essentially does not exist, apart from the very occasional short term individual. The 'housing first' policy ensures everyone has a legal right to a home, even if they have no money, have mental health issues or drug or alcohol dependency issues. In such cases the housing comes with a package of support to help with all these complex problems. It is acknowledged that while people are homeless and sleeping rough they cannot deal with these other problems, so providing a safe and secure home is the first point of support.

The American system seems to be like a factory production line to produce homelessness. Millions of people are working in low paid, short term, gig economy jobs. They are not in a position to reliably pay the rent every month. Added to this is the American health care system where sick people have to find the money to pay for their care. Many people with poor health find themselves uninsurable or inadequately insured and every year many such people in the USA are made homeless.

In Finland more secure housing tenure, higher minimum wages, free health care and welfare payments are all very much more generous, and of course all these factors help in preventing people becoming homeless. There is no factory production line producing a daily crop of newly homeless people as there is in the USA. Legal rights, taxation and well organised state provision seem to be effective in a way that any number of charities cannot ever be. The Finnish system seems to be geared toward a very supportive role for the state in helping all of its citizens. Anu Partanen refers to this as the 'Nordic Theory of Love'. The state sees its prime role as to be the loving development of all of its citizens to be the best they can be. The state offers substantial legal, financial, practical and moral support so that as many citizens as possible can achieve this. This is the model I draw upon with my proposed Global Green New Deal. The idea is to extrapolate this 'Nordic Theory of Love' into a 'Global Theory of Love' and to provide all the people of the entire world the same care and support currently flowing from the Finnish government to the Finnish people. To achieve this globally redistributive taxation is an absolute priority.

Philanthropy is a relatively weak force compared to taxation. Even Bill Gates, a great advocate of philanthropy, admitted that only about 15% of rich people give to charity. The vast majority do not. Philanthropy has never tackled big issues like homelessness or medical care effectively at a national or global scale. Given the current inadequacy of most governments it does of course play a vital role in trying to plug the massive gaps in government provision. Also given the vast inequalities in the world at the moment it is a way to for richer people to try to help the poor and needy. Some individual charitable projects are excellent in particular places and doing specific tasks, but it is no substitute for well organized structural change and comprehensive governmental support.

The proposals outlined in the Global Green New Deal would massively increase funding for health, education and other public services at a global scale. This would be predicated upon massive increases in taxation and a greatly decreased need for philanthropy. Also as the billionaire class would cease to exist one of the main sources of philanthropic giving would dry up.

Homelessness and inadequate housing are global issues. So if we had a form of global governance with good global scale leadership and a strong network of locally devolved governance structures, properly funded through globally redistributive taxation, we could effectively provide good housing to all 7.8 billion people. When we compared the factory production line producing an endless supply of vulnerable homeless people in the USA and showed how Finland had the policies that stopped this production of homelessness, it was then able to provide the legal right to housing and the practical help to make sure everybody actually had a roof over their heads. So key to solving global homelessness is restructuring the global economy to make sure that it does not endlessly generate more homeless people, and to do that all the aspects of legal rights, secure tenure, minimum wages, more secure employment, better systems of health and education are all issues that have to be tackled at the same time as more directly improving the supply of good quality affordable housing.

Replicating What Works

If we look at other issues we see over and over the repetition of failure. Take the war on drugs. Vast effort and resources have been put into this, yet almost nowhere has much progress been made. One country has taken a very different approach. Twenty years ago Portugal decriminalised the use of all drugs. Drug use became a health issue rather than a criminal justice issue. The number of drug related deaths fell and the proportion of the prison population with drug issues declined.[22] The funding that would have been used to fight a war on drugs was redirected towards counselling and other ways to help people make their lives more purposeful and healthy. Johann Hari's TED talk on drugs is well worth watching.[23] He locates the nature of addiction less in the chemical properties of the drugs themselves than in the emptiness and despair that makes people become addicted to drugs.

Tackling the epidemic of despair would seem a very good investment of resources. The proposed Global Green New Deal would have the universal provision of meaningful and purposeful work for everyone as a basic right, backed up by universal provision of counselling, health care, education, social support and safe and secure housing. The best bulwark against despair and futility is to find purpose and direction, and for loneliness and isolation to be replaced by being actively engaged with family, friends, communities and teams.

Very few countries have followed Portuguese model of drug policy, or the Finnish way of dealing with homelessness. The USA may be an extreme example, where market fundamentalism makes action on homelessness ineffectual and outdated moral concepts of punishment hamper effective drug policy. Instead of following models of success most countries follow models of failure. Outdated ideological attachments and shear inertia make progress slow or non-existent. Successful models and ideas about further improvements are often so far outside the Overton Window that they are never considered.

For example, the NHS is currently in crisis. Market fundamentalist ideology has been undermining the service with a goal of full privatization and adoption of an American system of health care. This would be a disaster. Opposition to this focuses on keeping the NHS as it was always designed to be, a National Health Service. The debate has become polarized. American style privatization would be the worst possible outcome, but keeping the NHS more or less as it is may not be the best. A careful study of better functioning health services might include looking at many European models, and perhaps those of Singapore, Japan and Korea, and thinking what is working best within those systems and how lessons could be applied to the NHS. Also within the NHS there are many areas of excellence, and some of failure.

What can be learnt and replicated from the Finnish policies on homelessness, Portuguese drug policy, Danish and German renewable energy co-operatives, or the health services of the dozen or so countries with the best health outcomes?

The Global Trust for People and Planet would be funding innumerable projects in our global network of local communities. Experimentation and the sharing of knowledge are critically important. We need the hive mind of humanity to be actively engaged in this whole process. We need to stop the endless repetition of failed policies. A liberated and happy workforce is essential to generate a plethora of projects and rapidly build on success across many fields.

Work and Consumption, Frugality and Liberation

Both capitalist and communist regimes have sought to control their populations and in many ways to limit the freedoms of their ordinary citizens, to keep them plugged in to 'the system'. Capitalism has used advertising to stimulate demand for more products than are truly needed, and hire purchase, mortgages and other forms of debt to keep people always in need of more money and always desperate for work. In the early days of colonialism local populations would often just take a job to get enough money to pay for a single item like a new axe and then give up the job and return to a self sustaining lifestyle. In Africa in the 1890s the colonial authorities brought in the hut tax specifically to create a need for money within the local populations so that they would remain in what were boring and exploitive systems of work.

Since the Second World War there has been a great focus on job creation. It drives economic growth, creating opportunities to invest and to make profits. It has been portrayed as the only pathway out of poverty and toward prosperity. Work has been portrayed as providing dignity and identity to whole communities. The wages generated by working created the demand that has fuelled these decades of consumer-driven capitalism. Economists portrayed human needs as infinite, so when we had a roof over our heads and enough food and clothing we were all encouraged to want more and better of everything, endlessly, forever. Advertising and the values of popular culture were used to reinforce these aspirational goals.

The cracks in the ideology and practice of consumer-driven capitalism have been apparent for many decades. Pollution, traffic congestion, environmental damage and habitat destruction were all getting worse during these decades of consumer-driven consumption. Many people did not experience the world of work as being one that generated dignity and identity but rather stress and frustration or boredom and alienation. Many also experienced the pursuit of endless amounts of stuff, from cars to televisions, ornamental vases to mobile phones, as pointless. A counter-cultural lifestyle emerged where it became almost a badge of honour not to buy new things but to reuse other people's cast-off items, so preventing these things going to landfill and saving money.

Mortgages, like all debt, keep us enslaved. Some of us took the struggle to live debt-free to extremes. I once lived in a garden shed while letting out my own house and working two jobs in order to reduce my mortgage and free myself of debt. In other phases of my life I often took off on long journeys, hitch-hiking, walking or cycling around the world. The goal was always to get to know and understand other people and other cultures and to develop friendly relations with people from as wide a range of backgrounds as possible. Living frugally, camping and spending as little money as possible were necessary in order to prolong and deepen the experience. The goal was the opposite of that portrayed by the values of consumer-driven capitalism. The goal was to explore a freer lifestyle, not constrained by the identities of workplace, profession, consumption patterns or even nationality or language. A personal sense of frugality became the path to liberation. We sought liberation from the drudgery of work and liberation from the endless need for more of everything.

Some of us in this movement drew upon various religious traditions and for some people the voluntary simplicity movement became a central theme of their lives. Pursuing more ethical patterns of consumption remains important for many people, and although companies try to exploit this to market more expensive products, striving for more ethical patterns of consumption remains an important trend. Most important is always to avoid the worst, and for many people now this means refusing to fly in aircraft, or to eat meat, or to buy products from countries with appalling human rights records. We each select criteria for judging what is worst and what is best, but there are many unifying themes among those of us pursuing this path. There is also the positive choice of where we do want to spend or invest our money, supporting ethical businesses, greener technologies, fair trade and organic and humane systems of farming. (We will explore many of these issues in the final chapter of this book, 'Creating Change: What you can do')

The challenge now is to imagine and create systems of employment that bring out the best of employment, of individual personal freedom and of the education and skills of the whole human population. Investing in the human population not to enhance the power of the state, or to generate markets for products or profits for corporations, but to be as capable, happy and fulfilled as possible and to be as actively engaged with working to heal the damage that has been done to people and planet under the current systems of economic and planetary management.

Reimagining Work

There has long been the problem of unemployment or underemployment. Over the coming decades it is predicted that automation will mean many millions more people will lose their jobs. Also as so much of our economy is geared toward creating jobs making stuff that soon ends in landfill, or working in industries that are polluting the planet, these unsustainable sectors of the economy will need to contract very quickly, thus making many millions more people unemployed. Two very different views are emerging. Those on the traditional left of politics tend to argue for the dignity of work and for full-time employment for everybody, and for them the prospect of rapidly decreasing employment is unacceptable. They often hang on to the notion that as human needs are infinite new jobs can be created indefinitely to cater for these infinite needs.

Whereas others, coming more from a greener perspective, tend to argue that human material needs are really limited, and that a reduced total amount of jobs in the economy should be a good thing. Why not bring in a universal basic income, a two-day working week so that the available work and income is more evenly divided, and also bring in long sabbaticals for those who want them, longer holidays and longer paid paternity and maternity leave. Working fewer hours in formal employment is important, so people have the time and space to develop more skills, hobbies and interests and to be good parents, and active citizens participating in building their local communities. Perhaps it is time that our dignity, identity, status and wealth should not be defined by accident of birth or by what we do to earn a living, but rather by how we participate in the life of the community, as parents and as citizens?

The dominant ideology currently assumes that the private sector alone can and should generate enough jobs to keep society functioning. However as we have seen during the decades of market fundamentalism, jobs have become less secure and lower paid for many millions of people, while for a few the rewards have become excessive. Reversing mass impoverishment will require strong governmental leadership. As many globally important companies effectively pay little or no tax the only way to make them pay their fair share is to introduce some form of global tax coordination or as I recommend, global-level tax collection. All tax havens and tax loopholes clearly need to be closed down.

Jobs always have been created outside the private sector. Governments have long raised armies and schools, hospitals and many other institutions which in most countries exist outside the private sector. Generally the more left-leaning governments were happy for a greater proportion of jobs to be in the state sector, and more right-wing, and especially those drawing on market fundamentalist principles, were highly suspicious of the state sector growing too big, with the one notable exception of the military. Conversely in communist countries any jobs outside the state sector were viewed with extreme suspicion. I am relatively agnostic about whether more jobs should be in the state or private sector, and there are advantages in having a good mix of the two. It does seem that some things work best the public sphere, such as health services and infrastructure such as the railways; and others, like shops and restaurants, flourish better as small locally-owned and controlled businesses. Many areas like the innovation of more sustainable systems of generating energy, or manufacturing better products, are usually initiated by individuals starting small businesses but requiring public sector agencies to help them during the initial process of bringing new products to market.

Given the scale of job losses due to automation, to the much-needed planned contraction of all polluting industries and the vast tasks ahead in building a socially just and ecologically sustainable global economy, some way of redeploying the human labour force is urgently needed. This transformation of the global economy gives us the opportunity to completely reinvent the daily pattern and experience of going to work. The important point is to make the work worthwhile, where it is helping to meet real human needs and helping to heal the planet, and where the daily experience of going to work should be filled with interest, enthusiasm and joy.

In the introduction I outlined a global agency that might be called the 'Global Trust for People and Planet' which could lead on this task. It would be funded by globally raised taxes and projects would be focused through our proposed global network of 39,000 or so local communities. To cope with mass unemployment in the 1930s Franklin D Roosevelt brought in the New Deal, creating many useful projects helping, for example, to heal the dust bowl by massive-scale tree planting. Roosevelt's Civilian Conservation Corps provides an interesting model. It had parallels with the Young Pioneer Corps in early Soviet Russia, and the Women's Land Army in wartime Britain or even the Scouts. All these organisations had the aim of building morale, physical health and skills in young people, and they also provided a ready-made workforce with a good team spirit, ready to get on and do a wide range of tasks. The proposed 'Global Trust for People and Planet' would share these goals but on a very much larger scale, globally organised, and offer much more comprehensive education and training opportunities, and much better pay and other benefits. It would also be the spawning ground for millions of small start-up businesses and co-operatives.

In thinking about how the world of work in general, and the proposed 'Global Trust for People and Planet' in particular, should be organised we have much to consider. Work has seldom been organised to bring out the best in people. It has almost always been exploitative, either in the interests of the employers and shareholders, or in the interests of the state. For many years I have had a lot of family and friends employed in the health services, mainly in the NHS, as doctors and nurses, paramedics, occupational therapists, and physiotherapists. The work is interesting, worthwhile and rewarding, yet most have left the NHS either taking early retirement or leaving to work in less well paid casual jobs, doing bits of self-employed jobbing gardening or the like. Asking them why they left the health services I almost always get the same answer. It is to do with the culture of stress and bullying that seems to flow from the top down through the organisation. Much of this results from successive governments making far too many reorganisations and changes in policy objectives and systems of measuring outcomes. Coupled with this have been long-term underfunding and excessive expectations. Departments are frequently short staffed and employees are then given impossible workloads, where they have insufficient time to really give patients the time and attention they deserve. This makes the work more stressful, accidents more likely to happen and poorer outcomes for both patients and staff. When health professionals leave the NHS they are often exhausted and demoralized. This is a tragedy. Things could, and should, be organised very much better, so staff are valued and nurtured. Rates of pay are one aspect, but far more important, and much less discussed, is how we could organize a health system, an education system, or any other area of employment so that the people working in it experienced less stress and more satisfaction. Obviously, from this would flow better outcomes for patients, schoolchildren and society in general.

Paul Verhaeghe is a Belgian professor of psychoanalysis and author, and in his book, *'What about me? The struggle for identity in a market-based society'*, he explores the impact of our current political and economic systems on mental health and the role stress in the workplace is having on employees and the wider society. The modern world uses almost exclusively extrinsic forms of motivation, such as money, praise or fame, to get people to work. However it is the intrinsic forms of motivation that are at the heart of work satisfaction. Intrinsic motivation is what drives us to do tasks without obvious external rewards, but rather things we do because we find them enjoyable, interesting, purposeful and rewarding. Could more work be along these lines? Verhaeghe and many others think so. Verhaeghe cites Dan Pink's focus on three key words: autonomy, mastery and purpose. Verhaeghe states 'Autonomy and mastery are closely related. Having a say over the organisation and content of one's work enormously increases motivation and commitment. This in turn leads to greater mastery and expertise, thereby increasing job satisfaction even more.'

Verhaeghe goes on to show how in our modern work practices, with huge organisations and little trust of employees' judgement, many people have responsibility without power. Dictates from on high must be delivered. This is a recipe for lack of job satisfaction. Verhaeghe, like Anu Partanen and dozens of other authors, give numerous examples of better systems of organising things. I want to take the ideas they present and apply them to my proposed Global Green New Deal.

Work and the Global Trust for People and Planet

My proposed Global Green New Deal envisages tens of trillions of pounds, dollars or Euros being raised through taxation and saved through stopping the perverse subsidies and waste of our present system. This money should be invested in tackling the vast tasks ahead of us. Hundreds of millions of jobs will disappear and perhaps a similar number, or maybe many more, could be created. All the problems associated with poverty and unemployment, as well as stress and burn-out stemming from poorly organised working practices, could be reduced or eliminated. We have briefly looked at how employment is currently organised and hinted at some better directions for employment policies.

We have explored the concept of moving beyond the nation state which currently divides the world into roughly 195 countries with very different levels of power, wealth and consumption patterns, and have proposed a network of 39,000 more or less equally sized local communities, all having somewhat more similar levels of power, wealth and consumption patterns. We have mentioned that these 39,000 communities would all have excellent schools and hospitals, housing and services, renewable energy co-operatives and farms supplying everyone with local, fresh organic food. International criminal courts would have very much more power to prosecute serious crimes, from tax evasion to organised crime, from genocide and crimes against humanity to ecocide and crimes against the planet. Global legal frameworks would ensure legal rights to breathe fresh air, to swim in clean rivers, for physical and financial safety, free health care and education, help, guidance and counselling, a right to universal basic income and a right to participate in meaningful and purposeful paid employment. With rights usually come responsibilities, and paying one's taxes, upholding the law and treating one's fellow human beings with respect would all of course be expected.

I have proposed a Global Trust for People and Planet as an agency working in all these 39,000 communities to make sure that the massive tasks of global transformation actually get done. I now want to explore how this Trust might also be a vehicle to pioneer new forms of employment. Paul Verhaeghe makes a strong case that job satisfaction is derived more from intrinsic rewards than extrinsic rewards. By introducing a universal basic income and universal rights to housing, and free education and health care, we would be reducing the extrinsic pressures to work. Most of the evidence stemming from the study of what makes human beings happy and fulfilled suggests that we all have a strong need to be actively engaged in meaningful tasks and projects, engaged in our communities and in doing useful work. Fostering and developing this intrinsic motivation to work would be a central theme of my proposed Global Green New Deal, and the formation of the Global Trust for People and Planet would be an excellent opportunity to experiment with new models of working.

I want now to explore the three key words identified by Verhaeghe and Pink: autonomy, mastery and purpose. Let us start with autonomy. In large organisations the power to make decisions is often taken by central governments or corporate headquarters and commands flow down to the workers who have the responsibility to deliver specified outcomes. This strips workers of autonomy, undermines motivation and leads to poorer outcomes. A better system would seek to empower the workers to develop their skills and trust them to design the best systems to do their jobs and to be supported where they need help. Anu Partanen shows how the Finnish education system achieves better results than those achieved in the UK or USA precisely because they have less top-down control and better trained and resourced teachers and schools, so both individual pupils and individual teachers have more autonomy in their work, more support and fewer dictates from on high. Power is decentralized, so everyone is more equally empowered, and therefore able to acquire higher levels of mastery.

Small organisations such as technology start-ups often have an extraordinarily enthusiastic atmosphere. This is especially the case where there is a strong belief in the social and environmental benefits of the product being developed. This enthusiasm is often linked to size and to autonomy. These start-ups tend to be small, with many important decisions made individually and collectively, with no dictates coming from on high. They exhibit lots of workplace autonomy. This sense of autonomy, of being in control, is closely linked to the collective spirit of enthusiasm. From this flows continual learning and improving one's skills and knowledge and consequent mastery over the destiny of the project, business or service.

Purpose is the third theme that Verhaeghe and Pink identify as crucial to a happy and fulfilling experience of the workplace. So much of our current economy has little or no purpose beyond trying to make a profit, and that alone is no route to a satisfying life. There must be a purpose to getting up and going to work, a sense that the result of one's labour is making things better for somebody. Working in education or in health care has the obvious positive outcomes in terms of children's learning and personal development or in treating an illness and seeing a patient recover. Craft traditions often engendered a sense of pride in a chair or cutlery well made or a house, a cathedral or a hospital well designed and built.

There is so much that needs to be done in the world to alleviate suffering and to ensure the reversing of the climatic, ecological and other many crises we face. There is no shortage of real meaning and purpose in doing these tasks. The aim of the Global Trust for People and Planet would be to create a workforce to perform these tasks and this massive workforce would be divided into many small teams with decentralized control and autonomy. Considerable time and money would be invested, continually enhancing the skills and abilities of these teams. With the autonomy that flows from this decentralized structure and the competence and mastery that flows from investing heavily in this workforce, and the huge importance of these many challenges and tasks, a sense of purpose should prevail. These factors should all make for strong intrinsic motivation and successful outcomes. There would be much trial and no doubt some error, but this looks like the best route to overall success.

Size and Structure

We are used to huge corporate structures dominating the economy. These are usually globally organised private sector companies. Some large nationalized industries have had similar structures, with heavy top-down management, such as the Indian Railway system or the UK's National Health Service. Size does seem to matter. In big top-down organisations the people making decisions almost never sit down to chat or to have lunch with the ordinary people working lower down in the organisations. Also pay differentials have become ever greater: the social and economic gulf between management and workers becomes ever greater. This is a recipe for poor working relations and lack of job satisfaction. There are alternatives.

Corporations can bring in workplace democracy and much more equal pay for all their employees, with chief executives receiving salaries and perks more or less the same as those of the ordinary employees. Ernest Bader and his wife Dora Scott founded the chemical company Scott Bader in 1921 and gave it to the employees in 1951, and they pioneered workplace democracy and more equal wages. It is still flourishing today and has spawned the charitable Scott Bader Foundation. In the USA Dan Price is the chief executive of Gravity Payments, and in 2015, very much against the grain of twenty-first century American business models, he introduced a radically flat wage structure with a minimum wage of $70,000. These examples of Scott Bader and Gravity Payments are the exceptions, and sadly the reality is that most corporations over the last few decades have become less democratic and wage gaps have grown ever wider. The Global Green New Deal proposals abolish any huge wage differences, setting global maximum and global minimum wages, and employing powerfully redistributive taxation to ensure ever increasing levels of equality.

Non-corporate.org is an organisation that promotes alternatives to the large corporate structures that have come to dominate the world. There are many types of organisations that fit into this non-corporate sector, including co-operatives, mutual societies, partnerships, employee-owned businesses, community supported schemes, trusts, social enterprises and not-for-profits, open source and freely shared technologies and many forms of self-employment. There are also alternatives like downshifting, where we reject the need for a product or service or DIY, where we do it for ourselves, whether this be growing our own vegetables or building our own houses. Throughout this book we are envisaging alternatives to the giant top-down hierarchies with their in-built inequalities and alienation from the decision-making processes.

In the chapter on politics we envisaged moving away from huge and unequal nation states to a networked structure of more or less equal communities. As we consider changes to the world of work it is possible to consider a similar shift, away from huge corporate structures, be they private companies or large nationalized industries, towards a vast network of smaller organisations. In organisations with millions of employees the chances are that the lowliest worker will never meet and feel comfortable sitting down to have a chat or lunch with the chief executive officer. In small start-ups such mingling is the norm. If we work in a workplace with up to a couple of hundred workers there is some chance that we will get to know most of their names and recognise them and be able to chat with them. Once an organisation gets any bigger than this, such personal connection becomes very much less likely. Bigger organisations are beginning to understand these dynamics and some are breaking their decision-making process away from top-down hierarchies to more decentralised and empowered teams, and in the process finding increased worker satisfaction and performance.

The envisaged Global Trust for People and Planet would be huge but far from monolithic or hierarchical. It would be much more a network of local organisations, with perhaps a relatively autonomous branch in each of the 39,000 communities. As with many universities it would seek to spin-out numerous start-ups, which might be not-for-profits or social enterprises, or self-employed sole traders, partnerships or co-operatives. There would be no overpaid and remote chief executive and management structure. Rather workplace democracy might decide to run citizens' assemblies of the workers to guide the overall policy and direction of the Trust. It would seek to utilize the hive mind of as many people as possible.

Let us return to the example of the UK's National Health Service. How would it be different if we applied the principles we have just been discussing? First, acknowledge that the basic concept of a health service that is free to use and accessible for all is an excellent one. The Covid pandemic has reinforced the madness of letting greedy government ministers issue contracts as a way of making private profits. The process of creeping privatisation should be reversed, and the need for any kind of private health insurance or private health provision should be reduced to zero.

The funds raised through taxation should go to the NHS directly, and a democratic structure made up of medical staff, patients and possibly some politicians or other representatives of the wider society, should set priorities and overall goals. Wages should be much more equal, so the lower paid would see generous increases and the more centralized upper management would see job losses and reduced wages. Funding should be increased and set so that long-term planning and investment is made easier. Individual hospitals and doctors' surgeries should have very much greater autonomy, and greater funding with plenty of help and support, but far fewer dictates coming from outside consultants or from government. Nobody should be expected to be working in under-strength teams where they are expected to do two or three people's workloads, as is now frequently the case. Within each of these hospitals and surgeries workplace democracy should be encouraged and team working developed. If the highest paid worker and the lowest paid worker in any organisation cannot, and do not, frequently meet and both feel comfortable in having friendly and constructive conversations about the organisation, then something is deeply wrong with that organisation. Sadly most organisations are like this in the UK and USA, but less so in Finland and the Nordic region, as Anu Partanen so vividly shows.

Transparency and accountability are also important. Currently many organisations persecute whistle-blowers, when they should be listening to them. A workplace culture that fostered greater equality ought to make this listening easier. The Global Trust for People and Planet would experiment with networked peer review structures where colleagues could democratically and openly discuss failings, and try to identify better practices or call in outside help. With greater autonomy and stronger teams of more equal employees a greater sense of trust in the team will emerge and a greater willingness to listen to ideas for improvement should prevail.

A Cohort of Young People

Currently about one hundred and forty million babies are born each year. Let us start by imagining the cohort born in 2005 who would now, in 2021, all be about sixteen years old. Let us imagine offering them all a new form of mixed work, education and training, like a ten-year-long apprenticeship. The aim would be to give them the very best education, training and practical experience of working in teams on real world projects undertaking all the necessary work to create the future we are envisaging.

Many of them would during these ten years of training have gained university degrees, carried out post graduate work, learnt new languages and experienced a rich mix of cultures, as well as developing a wide range of practical and social skills. They would be paid throughout, and at the end of ten years they would be offered further employment, or help to set up new businesses and projects. The aim would be to take a broad mix of young people from backgrounds as varied as possible and to bring out the best that they all can be. Probably not everyone would want to join such a programme, but given the tremendous opportunities it offers perhaps 50% of that year's cohort of young people would join. So if we had an annual intake of 70 million young people, once the project had been running for a decade we would have a pool of 700 million highly skilled and motivated people, aged between sixteen and twenty-six. These people would be working in projects in all our 39,000 communities. This workforce would represent pretty well all the cultures and languages of the world. All the participants would have the freedom to work in any projects anywhere in the world, and during their ten years of training most would work in a number of locations. These young people would be the foundation of the Global Trust for People and Planet and they would co-evolve with it.

Over the years I have visited a number of schools and universities as a student, or to lead workshops on possible solutions to climate change. When I was a student at the London School of Economics the student body was made up of 35% UK students and 65% students from abroad, and this seemed a very healthy mix. Years later I participated in a two-day workshop on climate change and possible action at Atlantic College in Llantwit Major in South Wales. It was one of the best organised conferences I'd ever been to, with dozens of sessions on diverse aspects of the issue. What was really impressive is that the whole thing was organised almost entirely by a small group of sixteen-year-olds.

Atlantic College offers the international Baccalaureate diploma and a good liberal education with real community service as a core part of the life of the college. It has about 350 students, aged from sixteen to nineteen, from 90 odd countries, some from wealthy backgrounds, some quite average and some from very poor or refugee backgrounds. They all seemed to mix well and to work well together. Atlantic College was founded in 1962 by Kurt Hahn and was the first of what is now a growing global network of eighteen United World Colleges, each in a different country. About 60% of the students at Atlantic College get some form of financial sponsorship, while the rest are fee paying. I wondered if this might be a model we could offer to all the young people of the world.

So, returning to our global cohort of 70 million sixteen-year-olds. We could offer them all an education starting off in many ways like that provided at Atlantic College. As about 350 students seems about the right size for such a college, which suggests about 200,000 colleges would be needed just to accommodate the sixteen-year-old cohort, so perhaps twice that to cover the two year international Baccalaureate. This would mean 400,000 colleges, which, spread between the 39,000 communities, would imply about ten colleges per community. That sounds about right. Young people might start off in their local college but move between colleges to gain different experiences. Of course we are not starting from scratch. In many places school sixth forms and colleges exist, and some no doubt are pretty good. However the scope for global scale expansion, improvement and international co-operation is massive.

At the end of these first two years most would continue their education and training within existing universities or in new institutions and projects, and during these last eight years of their training most would live, work and study in more than one country. During the whole ten years of training there would be a very wide focus on developing the whole person. All would receive a good broad liberal education, experience of workplace democracy and of working within communities to help alleviate real world problems and they would also naturally develop their career paths. As well as spending some time in established institutions studying things like medicine and engineering many would spend time in institutions that do not yet exist and that would be very much focused on global transformation.

We have looked a little at the significance of the epochal shift from 'The Fossil Fuel Age' to 'The Solar Age', and will return to this theme throughout the book. This transition will affect all aspects of life, and many new opportunities will emerge that require a lot of people with skills that as yet are little known. In each of the following chapters we will explore various aspects of this work. One theme that recurs in this book is that of the hot dry deserts becoming the epicentres for new forms of urban settlement based on the development of solar power. These new cities in the desert will require many thousands of solar engineers, and people with the architectural and building skills to design and build cities as radically more energy-efficient, more egalitarian, with zero use of fossil fuels, probably no private cars, and making great use of solar desalination, seawater greenhouses and other systems of growing food in the deserts. We will explore all this in later chapters. At this point I just want to highlight the connection of all this to our proposed massive cohort of young trainees. Many of them would spend time working in these hot deserts, developing the skills required to make these young people and these new settlements flourish.

In Chapter Five we will explore many of the roles that this cohort of young people might be engaged-in. For example I outline how solar energy and these new desert-based systems of land use might be used to meet a huge range of challenges. I outline how a network of projects might be based in the Middle East with the multiple objectives of bringing peace to the region, increasing economic, food and water security and generating vast amounts of zero carbon energy as well as sequestering massive amounts of carbon, so helping to mitigate climate change. This cohort of young people from an extraordinarily diverse range of backgrounds would have a vital role to play in active peace building. There would be the technical work of redesigning fractured infrastructure, of helping to heal traumatised individuals and communities and helping to establish new ecologically and economically regenerative businesses and projects, and to draw in and employ young people from all the former combat zones and help them integrate into well-functioning teams.

In areas of conflict, and in areas of acute poverty, providing excellent paid education, training and work along these lines would have several additional advantages. Young men are often lured into joining armed conflicts out of economic desperation. By taking most, or ideally all, of the young men from all sides in a conflict zone and including them in these multi-ethnic teams and giving them intensive counselling, guidance, education, training and work, as well as a good income, they would experience a better alternative to conflict. They would be constructing and staffing new teaching hospitals, schools and health centres, new systems of power generation and ecologically regenerative farming. The teams that they would be working in would seek to engender a sense of autonomy, mastery, purposefulness and that enthusiastic spirit that is so important to get all the necessary work done, and done well.

Chapter Three

Energy, Infrastructure & Materials

The Vision

As we make the transition out of 'The Fossil Fuel Age' and into 'The Solar Age' everything will need to change, not least in the ways in which we generate and use energy, the types of buildings and infrastructure we design and construct, and how our economy, and all of us individually, use 'stuff'.

The economy will need to switch from a fundamentally linear model to one that is overwhelming circular: from mine, use and throw away to continual reuse and recycling. We need to stop burning fossil fuels. To do this within the time frame that the climate crisis demands we need to do several things. The production of renewable forms of energy can be accelerated and we can reduce demand both by the adoption of more energy efficient technologies and also by changing lifestyles. As I have shown, massive political and economic changes will be required so that making the individual lifestyle choices that cause high carbon emissions or other forms of pollution become very much more difficult.

For too long these lifestyle changes have been characterized as 'hair shirt environmentalism' and as an endless imposition of limitations and restrictions. It is true that some activities and patterns of wealth, consumption and ownership will need to be restricted, but as I hope I have shown elsewhere, for the vast majority the kinds of changes I am proposing will involve a considerable increase in personal freedom, opportunity and quality of life, and indeed for most an increase in the standard of living generally. It is just that things will be different.

Constant change is one of the recurring themes of history. We are adaptable as a species. We have changed our ways and our technology many times before. But this time the changes will need to be more wide-ranging, more global and more rapid than ever before. We probably only have a very short timeframe before the climate and ecological emergency really spins out of our control. To quote the climate strikers' placards we must 'Act Now!' Similar economic turnarounds usually only happen in wartime; after Pearl Harbour the American economy changed in a matter of weeks to a war footing, and when Lloyd-George replaced Asquith in 1916 UK armaments production rapidly expanded. Peacetime transformations exist, such as Roosevelt's New Deal, or technologically, the rapid spread of computers and mobile phones. The transformation that is required is certainly larger than any of these examples, but humanity's technological and organizational potential is greater now than ever. The problem is political will; technologically the scale is large, but relatively straight-forward.

It is one of the key objectives of my proposed Global Green New Deal to get to 100% renewable forms of energy for all purposes in all countries and to do so as quickly as possible. This is a goal shared by a rapidly growing number of people and organizations all over the world. Getting to 100% renewables for everything will be technically challenging, but quite possible. Easiest, and perhaps most widely understood, is the shift from generating electricity from coal and gas to wind and solar. Changing the energy systems that power our transport systems, drive our industrial processes and heat our houses is much less well understood, and in some sectors is technically challenging. Some changes can be made very quickly and others more slowly. These changes will inevitably happen more quickly in some countries and in some sectors of the economy than in others.

In this chapter I shall outline some of the most promising technological aspects of this transition out of 'The Fossil Fuel Age' and into 'The Solar Age'. The envisaged Global Green New Deal sets out how this transition could be made in ways that promote global social justice and ecological regeneration at the same time as most effectively tackling the climate crisis. There are an extraordinary number of exciting projects around the world to learn from, to replicate and to expand. Some of these projects are concentrated on technical innovation, but the ones that I find most impressive are the ones that combine multiple objectives, for example generating zero carbon electricity while also achieving wonderful benefits for the local community and for the local wildlife. We have so many of these small scale projects to build upon. The Global Green New Deal seeks to tilt the field by changing the big picture of how the global economy works. At present it is so often tilted against what is ecologically sustainable and socially just. We need to think from the bottom up about how best to replicate these small but inspirational projects. We also need to think from the top down. We all live within global political, economic and legal systems that are preventing good outcomes. We need to think about what changes we would like at this global structural level that would best facilitate the kind of future we are advocating for.

At the heart of these changes would be a shift in how we all as individuals operate in the world. For centuries we have had massive power inequalities, systems of brutality and war, of divisions between imperial powers and colonized peoples, and in more recent decades the emergence of consumer capitalism and a rat race of competitive work, consumption and waste. My proposed Global Green New Deal envisages many changes intended to promote rapid reduction in carbon emissions, ecological renaissance and also, crucially, a shift to a system that promotes global solidarity and justice. That implies climate justice, social justice and economic justice. To do that we need to rethink how we ascribe value and ownership to resources.

Coal, oil and gas licences have been purchased from nation states by companies or operated by state-owned utilities. These fuels have been burnt and the fumes released into the environment with no thought for the effects. We now know the damage this causes. This implies we need to think differently about resources. If the rights of future generations to a stable climate, to fresh air and to unpolluted water are fundamentally important, then what right do national governments have to issue such licences? What right do companies have to profit from pollution? It seems natural to say nobody owns the wind or the sunshine, the oceans or the atmosphere. To many indigenous people the ownership of land was an anathema. Maybe the minerals within the Earth, and the soils and water of the Earth, just like the sun and the wind, should belong to everyone who is alive now, and who will be born in the years to come. As we use and reuse these resources the profits should be shared between everyone alive, and safeguarded so that the rights of future generations to make use of these same resources are protected. All these resources, like the oceans and the atmosphere, should be thought of as global commons. Improving the management of all these global commons underpins all that this book is about.

Building all the solar, wind and other necessary aspects of a renewable-energy powered global economy will require a lot of investment and material resources. Money and materials will need to be diverted from elsewhere. I have outlined how massive amounts of money have been wasted on perverse subsidies, quantitative easing, so called 'defence', and also on countless inappropriate infrastructure projects. I have also outlined how wealth and Pigovian taxes could be used to raise more funds. In the previous chapters we explored how many trillions of pounds, dollars or Euros could be saved and raised and I proposed investing this money principally through a global network of about 39,000 communities, with a strong global-level coordination and facilitation.

In Chapter One we looked at my envisaged Global Green New Deal. It had ten massively ambitious goals. Let me remind you of what they are:

1. To achieve net zero carbon emissions, globally and extremely rapidly.
2. To allow nature to heal and for biodiversity to flourish.
3. To end, or minimize, all forms of pollution.
4. To end poverty and inequality and the many associated social problems.
5. To end, or massively reduce, war and conflict.
6. To provide everyone on Earth with good food, housing and energy.
7. To provide everyone on Earth with free, excellent education and

health care.

8. To provide everyone on Earth with worthwhile work and economic security.

9. To improve physical and mental health, purposefulness and happiness.

10. To achieve a well-functioning, participatory, socially just global democracy.

Following on from these ten goals we looked at nine key areas of investment. Numbers one and two were:

1. Renewable energy systems to supply 100% of humanity's energy needs, including electricity, heating, cooling, industry and transport.

2. The establishment of a circular economy utilizing the very best cleantech to minimize pollution, materials throughput and energy consumption.

This chapter explores how best to achieve our ten goals within the context of these two areas of investment.

A recent trend is for large quantities of money, largely from the private sector, to be invested in wind, solar, green hydrogen and a range of other aspects of a cleantech economy. It makes financial sense already. Tragically governments, swayed by the vested interests of fossil fuel lobbyists, continue to subsidize fossil fuels, and this is slowing the rate of change, just when it needs to be increased. In some countries carbon taxes are slowly and tentatively being introduced, when it is obvious that they need to be introduced immediately, globally and boldly. These two aspects, the ending of fossil fuel subsidies and the introduction of carbon taxes are now fairly well understood as desirable policy objectives. Implementing them globally would tilt the field strongly towards low carbon technologies. That would be an excellent first step. The Global Green New Deal that I am envisaging would seek to start with these two factors, then rapidly build more levels of investment aimed at simultaneously achieving our ten outrageously ambitious objectives.

I have contrasted different economic models of ownership. The choice, from a British perspective, is often seen as between big private sector corporations and nationalized industries. In many other countries the co-operative and municipal sectors are very much larger. In this chapter I shall look at some of the best renewable energy co-operatives and municipal models of ownership and explore how they could be replicated and expanded into a global network reaching deep into the lives of people in all our 39,000 communities and be used as tools to achieve all of our desired goals. In the section on work we noted how in small start-up businesses there was often a passion and a spirit absent from larger and more established industries. We shall look at how this enthusiasm and creativity could be replicated and expanded so that many more people enjoy their work more, and also make a greater contribution to healing our many social and ecological problems. This chapter discusses energy and the following chapter investigates agriculture. In Chapter Five I shall build upon this and apply the thinking to various geographical locations.

Renewable forms of energy have a long history. The idea of these being the basis for 100% of all the energy used to drive a modern global economy is more recent, but has a longer history than many people realize. I shall now outline some of the key technologies and show how they could be put together in ways that achieve all of our desired goals. This implies big changes to systems of ownership as well as in technologies. In the final chapter I will investigate how we as individuals and communities can contribute to this process of global transformation.

Why Renewables? Historical Aspirations

In the late 1960s and early1970s I like many other people was deeply concerned about all kinds of pollution. Many groups pressing for more environmentally sustainable policies and technologies were founded in this decade, from Friends of the Earth and Greenpeace to the Ecology (later Green) Party and magazines like The Ecologist, Undercurrents, Resurgence and Vole all started publication at this time.

We were critical of pretty well all the ways electricity was generated.

Nuclear power had the dangers of waste, accidents and weapons proliferation. Coal, oil and gas contributed to global warming and created local air pollution. Big hydro dams often displaced many people and damaged river ecologies. Car use was increasing, and we knew the multiple levels of damage this created. Domestic heating was shifting from coal to gas in response to the Clean Air Acts, but this only reduced the pollution, rather than eliminating it.

Many of us tried to imagine and create lifestyles that were more sustainable. Using much less energy was often a major goal. I lived for months in my tent, and briefly in a cave, and stayed in various communes and intentional communities trying to pioneer more sustainable ways to live. For a while I volunteered at the Centre of Alternative Technology. We talked about, and experimented with, all sorts of renewable energy technologies. Even then it was clear that renewables were the safest and best path to a less polluting future. The vision many of us had was for humanity to be powered using only renewable energy. In those early days most of the technology was very small scale and the vision relied on people wanting simple low consumption lifestyles.

There had been earlier visions of moving out of fossil fuels and into renewables. In the late nineteenth and early twentieth centuries many experiments were made generating and using renewable forms of energy. I have previously mentioned pioneers of concentrating solar power such as Augustin Mouchot and Frank Shuman, and early discoveries in solar photovoltaics. In Denmark the first wind turbine to generate electricity was erected in 1891, and around this time hydro-electric systems began to get built. In the era before the First World War electric cars were as common as petrol cars, and both Thomas Edison and Henry Ford believed that electric cars were the future. In 1923 JBS Haldane gave a talk in Cambridge where he said that the UK's future energy needs would be met by wind-generated electricity and that this would be stored, via electrolysis, in the form of hydrogen. Some of these people active around one hundred years ago envisaged renewable forms of energy supplying the world's energy needs.

*

The Wasted Decades

Here we are in 2022, and we are in a climate and ecological emergency. So what happened? Why did we waste the last century? Many oil fields were discovered in the few years following the end of the First World War. The price of oil dropped and greater profits could be made from developing this resource rather than moving into renewables. Many of the renewable technologies were virtually forgotten. Then in 1973 came the first oil crisis. Prices shot up and many countries realized their economic vulnerability. A wave of interest in renewables started and many interesting projects emerged, but as soon as oil prices dropped again most governments and corporates lost interest. Only a very few countries sustained efforts to decrease their oil dependency, notable among them were Denmark and Sweden, but even their efforts were inadequate.

From early in the Industrial Revolution people understood the pollution caused by coal, and saw that eventually it would run out. In the nineteenth century there had been the discovery of the science of global warming due to carbon emissions, and I recall discussing this with my 'A' level geography teacher Tom Lancastle between 1971 and 1973: I gave a talk on it in 1972. I became fascinated by renewable energy and low consumption lifestyles around this time. I and the millions of others who shared my concerns and hopes were dismissed as crazy hippies by much of the mainstream. There were some wonderful projects to emerge at this time. In Wales the Centre for Alternative Technology, in North America the Rocky Mountain Institute and the New Alchemy Institute were projects that I and many others avidly followed. For years Andrew Warren and the Association for the Conservation of Energy lobbied government to take energy efficiency more seriously.

So much could have been done to use less energy and to switch from fossil fuels to renewables but tragically, in the main, governments were not listening. They were psychologically and economically embedded with the wasteful and polluting fossil fuel industries. Funding to research and implement a less polluting yet comfortable future was at best miniscule and patchy. Decade after decade was wasted. Between

1970 and January 2021 atmospheric carbon dioxide levels rose from 325.68 to 415.24 parts per million. We had ample opportunity to evade the worst of climate breakdown. At the fringes many of us advocated action. We wrote reports, gave talks, went on demonstrations, and some of us even invented better technologies.

Meanwhile the corporate world and governments did all they could to keep their business-as-usual model going. This included, in the case of much of the fossil fuel industry and the politicians they funded, systematic disinformation and outright lies about the severity of the unfolding crisis. The media, in the main, amplified this disinformation. The mega rich individuals who controlled so much of the global media were, and are, deeply embedded with the fossil fuel oligarchs and their pet politicians. Between them they managed to halt any effective action to reduce emissions.

However, despite all this opposition from the incumbent industries and a lack of political leadership, several positive developments did happen during these decades. Notably renewable technologies were improving, awareness was growing and change was happening at the fringes and in all sorts of positive ways.

The Reality Now

The last few decades have seen three powerful trends. The first has been increasing energy demand, due to the spread and intensification of consumer-driven capitalism in more and more of the world, coupled with increasing energy demand in the already rich parts of the world, and in middle income countries. Annual global primary energy consumption has nearly doubled from 87,599 Terawatt hours in 1980 to 171,240 in 2018.[24] Over this same period global population has risen from 4.5 billion to 7.8 billion, and this has been a factor in rising demand, but this should not be over-stated as much of this increase in population has been in those areas of the world with the smallest energy usage. Access to energy remains highly unequal, with just under one billion people, mainly in Africa, still living without electricity. Most of these billion people also do not own cars and have never flown in an airplane. They may have large families but their carbon emissions remain miniscule. By contrast there has been an exponential increase in the energy demands and carbon emissions of the richest billion people.

The second major trend has been the falling cost and increasing possibilities of a whole range of renewable energy and associated technologies. Over this same period, from 1980 to 2018, the cost of solar photovoltaic modules has fallen by 99%.[25] Even the International Energy Agency says solar power is now the cheapest form of electricity in history,[26] and predictions are for the costs of solar and other renewables to keep falling. Academics like Mark Z Jacobson point out the many advantages of switching the entire global economy to use 100% renewables energy for everything: electricity, heating, cooling, transport, industry and agriculture. Jacobson and others envisage global consumer-driven capitalism continuing to grow and spread around the world even as we quit fossil fuels, nuclear power and biomass. He argues that there are many benefits if we can make the transition to just solar, wind and water (the term water here covers hydro, pumped hydro, tidal and wave power).

The third major trend has been the unfolding climate and ecological crisis itself, and the popular protest movement to reverse all the policies that have allowed this planetary-scale catastrophe to develop. Evidence of the sheer scale of the destruction mounts daily. The scientific data is overwhelming. The number of climate-related humanitarian tragedies increases year by year. People around the world are rising up in protest. Greta Thunberg has inspired millions of people to join in school strikes, go on demonstrations and lobby politicians. Greener political parties are slowly gaining in popularity. The question many now ask is 'Have we left it all too late?' Had we started on a path to a low carbon economy back in the early 1970s when atmospheric carbon dioxide levels were just 325 parts per million we could have made a slow and gentle transition. For most of the last ten thousand years levels fluctuated a bit either side of 285ppm, and 325ppm was still under the generally agreed safe upper limit of 350ppm. Now we are at 415.24ppm and in the danger zone, and levels are still rising. There is a sense of urgency: rapid bold action is required now.

To me, and to most people, it is clear we need to reduce emissions as fast as possible. Within the energy sector we need to push ahead on two fronts simultaneously. We need to invest massively in a raft of renewable energy and related technology, just as Jacobson and others advocate. We also need to vigorously pursue policies to dampen down demand. There are two aspects to this demand reduction. One is that by investing in very much more efficient technology we can save energy. The second strand of demand reduction focuses on political and psychological aspects rather than just the technological. This pathway draws on a very different tradition of thought, one advocating the benefits of living a simpler life deeply embedded in strong communities in a peaceful and just global system. This implies massive scale global redistribution of wealth and power, and a need to focus investment in the public realm rather than the private.

My envisaged Global Green New Deal would seek to push both aspects of demand reduction, so that overall global primary energy use stabilizes, or even falls, at the same time as renewables are quickly ramped up, and global access to energy and to wealth becomes very much more equitable.

Clearly we should have done more to reduce emissions many years ago. However things are not beyond hope. Powerful trends are on our side. First the renewable energy technologies are much better and cheaper than they have ever been: technically we could decarbonise the global economy and provide a reasonable standard of living for all 7.8 billion of us. Economically a tectonic shift is occurring: the share price of many coal, oil and gas companies is falling and that of renewable energy companies is rising. Economic and political power is moving from those blocking action to those advocating action. Although in many countries the politicians so embedded in the fossil fuel era mindset are still in power, the forces opposing them are growing stronger. Politically we still are a long way from where we need to be to implement any policies likely to reverse the catastrophe, but political change can happen very suddenly in all sorts of ways. Biden replacing Trump may be one tiny step in the right direction. More Green politicians being elected to local governments across the UK, and into national parliaments across the EU, is a bigger step. Extinction Rebellion, the global school strikes movement and the pro-democracy protest movements all over the world are vital parts of the pressure to move in the right direction. None of these things is going to create the necessary change, but they may just be small parts of a very much larger cascade of changes about to sweep the planet (as we saw in the chapter on politics).

As I write this in November 2021 the COP26 climate conference in Glasgow is drawing to a close. The final communiqué looks very weak. The number of fossil fuel company delegates was greater than the number of delegates from any one country. The UK budget the week before COP was a case study in what not to do. The UK government under Boris Johnson is the worst in living memory, up to its neck in corruption, without a moral compass and with no sense of seriousness about reducing carbon emissions.

But in Glasgow there were many signs of hope. The Beyond Oil and Gas Alliance has been formed, and is being led unsurprisingly by Denmark and Costa Rica, with France, Greenland, Ireland, Quebec, Sweden and Wales joining and New Zealand, California, Italy and Scotland showing support. Inside the conference halls the best speeches were from Mia Mottley the prime minister of Barbados, and Liz Wathuti, a Kenyan climate activist and founder of the Green Generation Initiative. Out on the streets of Glasgow there were huge demonstrations with hundreds of thousands of activists and eloquent leadership provided by Greta Thunberg and Vanessa Nakate. Behind the scenes there was much useful networking among cleantech businesses, activists and many others. I think it can be said that overall there was a lack of high level leadership from the big countries, but lots of positive things happening on the fringes, and to some extent in the small technical details that are being negotiated.

One indication of how the UK government is still in the hands of the big incumbent industries, especially the fossil fuel and nuclear industries, is its failure to fund research into models of how the UK could achieve 100% renewable energy use across all sectors of the economy. A group of academics and enthusiasts formed the organization '100% Renewable UK' and hoped to commission some research led by Christian Breyer and his team from LUT university in Finland.[27] To do this they raised money from the public, with Greenpeace contributing, but nothing from our government. Any government serious about carbon reductions would be keen to fund and to listen to the findings of such research. Sadly this does not seem to be the case with our current government. Imagine how different things would have been if Caroline Lucas had been leading a Green government instead of Boris Johnson's Tory government. The 100% renewables research would have been funded, listened to and acted upon and the COP conference in Glasgow would have been very much more successful. Something like my envisaged Global Green New Deal might even have been on the table!

Waste and the Circular Economy

The various stages of 'The Fossil Fuel Age' have been in large part symbolized by the changing nature of our waste streams. The dark satanic mills of the nineteenth century belched dark smoke from their chimneys. Domestic heating and the steam railways also burnt coal. As the air quality in cities became unbreathable during the pea soup smogs of the 1950s the response was to bring in Clean Air Acts banning the domestic burning of coal in cities such as London. Instead coal was burnt in power stations to make electricity, and the power stations had very tall chimneys. The pollution was shifted higher up into the atmosphere. This improved air quality in London, but masked rather than cured the pollution. The pollutants were shifted elsewhere, not eliminated. The so-called solution to air pollution allowed problems such as acid rain and climate change to worsen.

As the twentieth century developed so too did consumer society and the throw-away culture. Our domestic waste streams grew ever greater in volume and contained an ever increasing number of different pollutants. As the human population grew and consumer capitalism spread around the world, the mountains of rubbish grew ever bigger. We all threw things away. The local council was then responsible for dealing with these increasing amounts of waste, which mainly went to landfill. As landfill sites became harder to find, incineration and recycling became two possible alternatives, but neither is without problems. Incineration has to be done well in very high specification plants to make sure that all the toxins such as dioxins are properly burnt, and the public needs to trust the emission monitoring process. Even if it is done as perfectly as possible it still does nothing to reduce the sheer volume of materials flowing through our economy, with all the associated energy use and pollution that this entails. Recycling has been held up as a kind of panacea, yet time and again we have found that domestic recycling ends up being exported to countries such as China, Turkey or Malaysia where it is often illegally dumped into the environment.

Over the last hundred years as petrol and diesel engines were developed, roads began to criss-cross continents, and the numbers of cars and lorries rose exponentially. Shipping and aircraft proliferated. All of these transport systems burnt fossil fuels and emitted a range of pollutants.

All of this history of the waste and pollution that humanity has generated during 'The Fossil Fuel Age' has been predicated upon the belief in 'away'. We could throw away our old televisions, mobile phones or plastic bags, and our factory chimneys, power stations, cars and planes could emit smoke and pollution into 'the environment', where it was supposed to dissipate, to be blown 'away'. Now we understand that we live on a finite planet, and everyone's 'away' is someone else's lungs and livers. Every human being on Earth now contains the traces of a horrifying range of pollutants. Echoes of the atmospheric testing of nuclear bombs from the 1950s and of DDT used in the 1960s still linger in the bodies of children born today. Microplastic particles are to be found everywhere, including inside the bodies of everyone on Earth.

We know for example that sperm counts are falling, certain types of cancer are rising, and anxiety, depression and other types of mental illness are becoming more common. We also know that many carcinogenic and endocrine-disrupting chemicals are widely sold, and also that modern life is fraught with many stress factors. The causes of most individuals' illnesses will be multi-factorial. In myriad ways we are undermining the health of ourselves as individuals and also we are undermining the viability of the entire biosphere. As we encroach ever deeper into the natural world and continue to disrupt the global climate we make pandemics such as Covid 19 and SARS ever more likely, and as we fly around the world we ensure the rapid spread of these new diseases.

Climate and ecological activists are rising up, and one of their slogans is "We are Nature, rising up to protect itself". Can we really imagine a system that did put the health of all humans and of the entire biosphere above the right to make profits or to seek power? That might entail an end of the concept of 'away'. Could a society be created that threw as close to nothing away as possible? Might that mean a radical reduction in what could be bought and sold? There are probably hundreds of thousands, or maybe millions, of products that should simply be banned. The global economy would need to shift from an economy that was linear to one that was circular.

Again, the climate protestors have tried to sum up these complex changes in production and consumption in slogans that will fit on to a placard to be carried through the streets. One said "If it can't be reduced, reused, repaired, rebuilt, refurbished, refinished, resold, recycled or composted, then it should be restricted, redesigned or removed from production." Well, that does just about sums it up.

Renewable forms of energy have the obvious advantages of not creating the air pollution and carbon emissions associated with burning fossil fuels, or the risks associated with nuclear. It is of course imperative that as we try and halt the damage done by burning fossil fuels we do not create new problems of toxic waste streams. Currently one of the downsides of the shift to renewable energy is that old wind turbine blades, solar panels and many other vital parts of a renewable energy system such as batteries have been sent to landfill. This is beginning to change. As all these things become more common, and more of them come to the end of their useful lives, it has become commercially viable to initiate recycling of a growing number of the components. Of course funding and legislation should have been in place to ensure that this recycling replaced wastefulness earlier in the cycle of product development and deployment, and with the kinds of changes to tax and investment outlined in my envisaged Global Green New Deal this could have been the case.

Global Renewables

Many people still do not believe it is possible to have a reliable electricity supply from just wind and solar power. Obviously if we think in terms of a single solar panel it will not be generating electricity in the hours of darkness, nor will a wind turbine be much use in very still weather. However these problems are rapidly being overcome by an evolving mixture of grid interconnections and various methods of storing energy. So if one thinks of a globally integrated system, the wind is always blowing and the sun is always shining and the potential global energy resource is relatively constant and abundant. These comments about a single wind turbine or a globally integrated electricity grid are the theoretical extremes. In reality energy systems are rapidly changing, systems of energy storage are being developed and grid interconnections are linking up national grids, and all this is making wind and solar a realistic basis for most countries' electricity supply and indeed for all other uses of energy such as heating and cooling, transport and industry.

Some sectors like steel making, air transport and certain plastics do present challenges, but eventually they too will be based on renewable resources. A few activities like air travel will need to be rationed until the technology becomes very much less damaging. Increasing the share of renewables on the electricity grid is the first step, and many countries are well advanced with this. About a dozen countries already have 100%, or very nearly 100%, renewable electricity. They are mainly fairly sparsely populated and have a lot of hydro, such as Paraguay,

Bhutan and Norway. New Zealand, Iceland and more recently Kenya have added significant geothermal to the mix. Costa Rica is pretty well 100% renewable with hydro, geothermal and some wind and solar. Scotland, Denmark and Ireland are three of the first countries where wind dominates the grid, although none is yet using 100% renewables, but they soon will be. None of these countries is very densely populated or heavily industrialized.

Germany is famous for its 'Energiewende' or renewable energy transition. It must be one of the hardest countries in which to make this transition, being densely populated and heavily industrialized and with a relatively poor wind and solar resource. Germany will struggle to get to 100% renewables from just its own local wind, water and solar. It is doing all it can to develop these local resources, but it will probably be a net energy importer of renewables, just as it is now for oil and gas.

Some countries have vast renewable energy resources, far greater than their own energy demands will ever be. There will be long distance trade in renewables just as there is for fossil fuels. In Europe Norway, Sweden, Iceland and Scotland will be net energy exporters, with their huge wind, hydro and, in Iceland's case, geothermal resources. Germany and Poland will be net energy importers. Electrical grids are expanding with high voltage direct current cables efficiently bringing electricity from areas of easy surplus to areas of greatest demand. Such electrical interconnectors have been around for decades and they are currently going through a rapid process of expansion. The vast solar resource of North Africa will be developed both for increased local need and to export energy to power-hungry Europe. More electric cables are being planned to cross the Mediterranean. However there are other alternatives for the long distance transport of energy, and chief among them is turning surplus renewable electricity into hydrogen and shipping it or sending it through pipelines to countries of high demand. Hydrogen can also be converted into ammonia, methanol or other fuels which can then be used directly or converted back to hydrogen, or to electricity.

In December 2021 Siemens Gamesa signed a memorandum of

understanding with Strohm to jointly develop thermoplastic composite pipes.[28] The intention is to locate electrolysers on North Sea wind turbine towers and send hydrogen by pipelines across the sea floor back to Germany. This may well prove cheaper than electric cables, and it would add a very useful element of energy storage to the grid. I would expect a lot of energy to be transported this way within a few years.

Australia is now positioning itself as a major exporter of hydrogen, utilizing its vast solar and wind resource to supply power-hungry counties like Singapore, Japan and Germany. [29] In the case of Germany it will be hydrogen, methanol or ammonia by ship. It also plans to export electricity via cable to Singapore, linking it into the emerging sixteen nation ASEAN power grid. Singapore currently has very high per capita carbon emissions and with its tiny size, very limited resources, dense population and high level of industrialization it has to import energy, now from fossil fuels and soon from renewables.

Mike Cannon-Brookes, the Australian tech billionaire, is part of a consortium proposing to build what would currently be the world's biggest solar farm on 15,000 hectares of land at Tennant Creek in the Northern Territories. He is planning a 10GW solar farm, including 22GWh of storage and a 3 GigaWatt, 4,500 kilometre cable linking to Singapore, all at a cost of $22 billion.[30] This cable would supply 20% of Singapore's electricity needs. Plans are already underway to build the solar, wind and battery system, and the interconnector, by 2027. This first cable is just the initial step for a country like Singapore in its transition to a renewables-based economy.

Mike Cannon-Brookes thinks this will be the first of many export cables for Australian renewable electricity. He also points out the advantages of running cables across Australia in an east-west orientation. This would mean Australia could switch to 100% renewables with massively decreased need for energy storage. About 75% of Australia's population lives on the east coast, and during their peak electricity demand from 5pm to 9pm they could be powered directly from Western Australia's sunshine due to the three to four hour time difference across the country. In Europe the power grid is expanding and linking together wind farms on Ireland's Atlantic coast to the North Sea and all the way east to the Baltic, supplying electricity into most of the countries of northern Europe. This wide range of interconnection has the advantage of helping to smooth out the fluctuating nature of wind power, and supplying the electricity to wherever the demand is greatest at any given time. Within a decade or so much of this long distance trade in energy might be hydrogen via pipe rather than electricity via cable.

As we move from 'The Fossil Fuel Age' into 'The Solar Age' long distance trade in energy will continue, but instead of oil, coal and gas, it will be of electricity and hydrogen and a number of other renewable-energy storage gases and liquids such as methanol and ammonia. Some countries, like Saudi Arabia and Australia, which are currently big exporters of fossil fuels, will switch to being big exporters of renewables. Early signs of this are emerging in Australia.

Morocco and Chile are two countries with few fossil fuel resources, but vast renewable potential which they are just beginning to develop. I think it likely that these will be two of the biggest net energy exporters in years to come. Morocco already has a grid interconnection to Spain, and so into the rest of Europe. More cables will be laid across the Mediterranean.

Chile's Atacama Desert is relatively high altitude and dust free with strong year-round sunshine which means that more solar energy can be generated per square metre than anywhere else on Earth. Therefore Chile should be able to produce solar energy a bit more cheaply than anywhere else on the planet. Chile is just beginning to develop this vast resource and has plans for massive expansion including for the large scale export of green hydrogen. Chile's Tierra del Fuego is one of the best locations in the world for developing wind, and with minimal local population or energy demand this could add significantly to Chile's energy export potential.

Just as coal transformed places like Germany's Ruhrgebiet and many of the towns of northern England in the nineteenth century, so solar and wind power will have utterly transformative effects on certain places on the planet. Just as in the nineteenth century, investment and people will flow to where energy is cheapest. Power-hungry heavy industry, including for example aluminium smelting and steel making, tends to move to where energy and resources are cheapest. In the nineteenth century this was to the coalfields; in the twenty-first century it will be predominantly to the hot dry sunny deserts of the world. As industries move it opens up other opportunities. New technologies like solar desalination are creating possibilities for new forms of human settlements in the hot dry deserts of the world. I will explore this further in chapter five.

Storage

We have seen how expanding the interconnections between national grids helps make the sun and wind much more useful as the basis for a reliable energy supply. Another factor co-evolving with grids is energy storage. There are a bewildering number of ideas and technologies that are part of the energy storage picture.

Batteries are by far the best-known devices for storing electricity. We all have them in many devices. They are very useful for storing small amounts of electricity for short periods. Their falling price and improving performance means batteries are increasingly being used for load balancing on electricity grids and in electric cars. However other systems of energy storage are more useful for storing larger amounts of energy for longer periods. Pumped storage hydro, such as at Dinorwig in North Wales, are very useful, and more pumped hydro will be built where the geography is suitable. There are dozens of other methods of storing energy being developed. Some use gravity, as pumped storage does, but with concrete or other weights instead of water. Gravitricity[31] are currently trialling a system in Leith, Scotland, that they then hope to deploy, lowering weights down old mine shafts to generate electricity at times of peak demand, and using electricity at times of peak supply to haul the weights back up. Other people are developing storing energy using compressed air, flywheels, phase-change chemicals and many other methods.

Using electricity to divide water into oxygen and hydrogen via electrolysis has been known about for over a century, and much has been written about the potential role of hydrogen. Suddenly in the last year or two investments have flowed into the sector and numerous projects are being developed and built. Incumbent industries such as coal, gas and nuclear are all talking about hydrogen production as a way to keep their struggling industries afloat. As solar and wind costs fall and the technology is inherently less polluting, making so-called green hydrogen from surplus renewables seems to be the way to go, and this is where the biggest investments are happening. In the section on transport I shall discuss the relative benefits of hydrogen versus batteries. We do need both as they each have somewhat different advantages.

There are a number of ways of using heat and cold to store energy. With concentrating solar power it is standard practice to store heat, usually in the form of molten salt, to be used to generate electricity after the sun has set. This is being developed in hot countries such as Spain and Morocco. District heating systems, such as in Denmark, often have large volumes of water acting as interseasonal heat stores. In windy weather surplus electricity can be dumped into these heat stores to be used many months later. Conversely surplus electricity can be used to make ice, which can be easily stored for months and be used, for example, in district cooling systems. This has huge potential for replacing current systems of air conditioning which are wasteful of energy and sources of pollution. Storing surplus renewably generated electricity as heat or coldness has tremendous potential.

Solar Technologies

There is a range of solar technologies, and by far the best known is the solar photovoltaic panel. The first practical application of these panels was in 1958 on the Vanguard 1 satellite. At that time they were extraordinarily expensive. Prices have come down in a remarkably consistent way. With each doubling of installed capacity the price of solar modules has dropped on average by 20.2%. So between 1976 and 2019 the price of solar modules has fallen by 99.6%. The fall in the last decade has been 89%.[32] The energy required to make each panel has fallen as new materials and methods of production have been applied. Electricity from solar photovoltaic panels is now the cheapest form of electricity in many places in the world. The price of solar panels is absolutely certain to continue to fall as new materials such as graphene come on stream. Gradually as more solar panels reach the end of their lives more and more will be recycled with the constituent parts made into new panels, or other products. Solar photovoltaic panels have the one obvious disadvantage, in that they only produce electricity when there is sunlight.

Concentrating solar thermal power is an older but a less well known technology that nowadays is usually built with thermal energy storage as an integral part of the system, so electricity can be generated after the sun has set. Such plants can if necessary be designed to run twenty-four hours a day, but more often just during the day and for a few hours in the evening and early morning periods of peak demand. Mirrors and lenses are used to focus the heat of the sun on to a heat transfer fluid, or directly on to a boiler, to generate steam and drive a turbine in much the same way as nuclear, coal or gas power stations generate heat to make steam and drive a turbine. I have been a supporter of concentrating solar power for decades and have blogged about it frequently.[33] There are many designs, with different configurations of mirrors and lenses. Parabolic troughs are currently the most common, but heliostats focusing the sun's energy on to a central power tower are of growing significance. Other designs, such as the compact linear Fresnel reflector or the parabolic dish are also sometimes used.

The fact that concentrating solar power systems use heat storage so that they can generate electricity after the sun has set is a huge advantage over photovoltaic panels which would need batteries to add storage. As heat storage works better than batteries to store large quantities of energy for longer periods, I think that this will become increasingly important as we move toward a global system of 100% renewable energy. There are also many different methods of storing the heat, with molten salt currently the most common, but pebbles, molten aluminium, ceramic particles, air, graphite, tanks of hot water or pressurized steam are all heat storage media I've read about, and I am sure there are many more, most at the research and development scale.

Concentrating solar thermal systems work best in hot dry climates, whereas solar photovoltaic panels are much better suited to a wider range of climates. The dramatically falling price of photovoltaic panels has meant that to some extent they have eclipsed solar thermal, but solar thermal has several advantages, notably in generating electricity after the sun has set, as discussed. And it is a more flexible technology with many potential uses apart from generating electricity. It too is going through a process of rapid cost reductions, just not quite as quickly as photovoltaics.

Concentrating solar thermal systems can be used for a great many other functions apart from generating electricity. The heat that the mirrors and lenses create by focusing the sun's energy can be used directly to smelt metals, or drive all manner of industrial processes, almost all of which are at the embryonic stage, but which have huge potential. Airlight Energy's giant solar collector at Ait Baha in Morocco uses a huge concrete parabolic trough to heat air that is used in an adjacent cement factory.[34] A pasta factory in southern Italy is planning to build a solar power tower to provide both heat and electricity to be used in the factory.[35]

Mirrors and lenses can also be used in conjunction with photovoltaic cells in a technology known as concentrating photovoltaics. A few years ago there was tremendous excitement and many new start-ups in this field, but the falling cost of standard photovoltaic panels has somewhat eclipsed this technology.

Another old and relatively simple technology is the solar hot water panel, which is best known in its simplest form for domestic hot water. In some countries, such as Turkey, these panels have been very widely deployed. In Denmark most houses are connected to district heating, and sometimes district cooling networks, with these types of solar panels feeding hot water into the system. Such solar hot water panels can be used to cool buildings, via the use of an absorption chiller, as pioneered by the Austrian company S.O.L.I.D. solar.[36] Solar power can also be used, either directly as heat or as electricity, to desalinate sea water, and solar powered desalination will be of critical importance as many parts of the world face a crisis of water supply, and as predicted in this book, new cities grow in the deserts.

There is a lot of exciting research and development, and many new start-up companies in the whole solar energy sector. Some of these technologies, and combinations of technologies, have tremendous potential. One of my favourite companies is Naked Energy which combines solar water heating and photovoltaics into a single panel. The water cools the photovoltaic cells which means they operate more efficiently, which will be especially beneficial in hotter climates. By also producing hot water from the same panel it means that they collect more energy per square metre than ordinary solar panels.

Solar is currently the fastest growing energy sector in the world. Under my envisaged Global Green New Deal this would accelerate exponentially. Where to place all these solar panels is an important factor to get right. Rooftops are the obvious place for household-scale production, and factory roofs make great locations for larger units, but we will need very much more than just rooftops. A small American start-up called Solar Roadways tried to develop a road surface that integrated solar cells, and various other projects have explored similar applications, but so far none has been sufficiently cost-effective or durable enough to withstand traffic wear and tear. Some ordinary solar photovoltaic panel systems have been built in ways that have beneficial effects, for example by deploying floating solar arrays over reservoirs or building them over canals in hot arid climates, rates of evaporation and water loss can be greatly reduced, and the waters' cooling effect adds to the panels' efficiency.

Most large solar arrays are field scale. One criticism of renewables is that if deployed at scale they can compete with other land uses, and it is true that much solar power has been deployed in ways that detract from existing land uses, especially agricultural production, and sometimes also their deployment has resulted in decreased biodiversity. The emerging field of agrivoltaics offers the prospect of best combining solar arrays with other land uses, and the potential for mutual benefit is huge.

Agrivoltaics

The term agrivoltaics combines the words agriculture and photovoltaics. If done well many benefits can be achieved, from biodiversity gains to more productive farming systems. Early research on how best to combine agricultural production with the deployment of solar arrays was carried out by the Fraunhofer Institute in Germany. In Japan a thousand or so projects are already up and running. In most countries this concept of optimising multiple benefits is still relatively unknown. In the UK a number of field-scale solar arrays are being developed in conjunction with local wildlife trusts in order to establish wildflower-rich meadows, attracting bees, other pollinators and insects and so also birds, bats and other predators. As far as I am aware none yet in the UK has the goal of also adding additional benefits from developing new and more productive agricultural practices. This will soon change.

One of my favourite projects is in the American southwest, the new Colorado Agrivoltaic Learning Centre[37], located on Jack's Solar Garden. This is a five acre solar farm on Byron Kominek's family farm, 24 acres on the outskirts of the city of Boulder, Colorado. Here they do research and educational work with the intention of getting other farms in the area to adopt these optimized combinations of solar power with other land uses. The solar panels provide protection for the fruit and vegetable crops during hailstorms or other severe weather. They also help reduce the extremes of heat and cold, reducing the stresses that these cause to crops. One of the chief advantages in the hot dry climate of the American southwest is that the crops need less irrigation due to the shading provided by the panels. Byron Kominek has a board of experts to help study what works best in terms of both agricultural production and biodiversity gains. By incorporating many species of plant native to the area they hope to increase local insect and bird populations. He also employs an educational and outreach assistant and they welcome school, college and other groups on tours.

This project in Colorado, like nearly all agrivoltaic systems, uses standard photovoltaic panels. German renewable energy company BayWa and the Dutch company Groen Leven have co-developed photovoltaic panels with various degrees of transparency, optimized for growing different crops and for different climates. In 2020 they started trials with various types of soft fruit at five test sites in Holland.[38] Growing fruit and vegetables under solar panels has many advantages. In hot dry climates water saving is a major factor. In cooler climates protection from frost might be a bigger factor. The solar arrays structure can be incorporated into supports for crops like grapes, climbing beans or raspberries. I think it very likely that the kinds of clear photovoltaic panels currently being developed by BayWa and Groen Leven will replace existing polytunnels and be used in greenhouses in many locations all over the world.

I know people have discussed combining concentrating solar thermal with farming, but I am not aware of anyone yet doing this in practice. These various forms of solar power, when optimally combined with food production, have enormous potential.

In my envisaged Global Green New Deal a large proportion of the world's energy supply would come from this as yet tiny sector. I see it as being transformative in both the global energy supply, and the global food supply. The Colorado Agrivoltaic Learning Centre is one of those inspirational projects that I would like to see replicated. We could think of using the Global Green New Deal to set up similar projects in most climatic zones on Earth, trialling different crops on different soils in different climates. We might want hundreds of thousands of such projects, some on a very much larger scale. I would love to see large scale systems trialling agroecological polycropping systems co-developed with large scale solar systems utilizing the whole range of solar technologies.

By far the biggest agrivoltaic systems anywhere on Earth are now being developed in China's Gobi Desert. One such specializes in growing goji berries under a 1GW solar array. Others are experimenting with a great range of fruit, vegetables, mushrooms and aquaculture under the solar arrays.[39]

In the following chapter I shall talk more about how agrivoltaics could be used in relation to food production and in Chapter Five I look at the potential in relation to the Global Green New Deal and how it might be developed in specific climates and locations around the world from the perspective of maximizing a whole raft of benefits. Suffice to say in this chapter on energy production that a very large share of the global electricity supply could usefully be generated in agrivoltaic systems.

Wind

The price of wind generated electricity continues to fall as the technology improves and even in the existing economic system it is outcompeting fossil fuels. The falling price of wind power has not been quite as rapid as photovoltaic solar, but still impressive. With the kinds of changes envisaged in this book the global adoption of wind power would be massively accelerated. Perhaps not to quite the same extent as solar energy on a global scale, but for the windy and not very sunny parts of the world, such as the British Isles, it would become the major source of energy. I have frequently blogged about wind power.[40] The size of individual turbines keeps increasing, as does the size of new windfarms. Improving grid interconnection and various systems of energy storage are all making wind power more useful and more reliable. Haldane's prediction from 1923 that the future energy driving the UK economy would be in the form of hydrogen derived via electrolysis from wind generated electricity now looks close to rapidly, and belatedly, becoming reality.

The shallow waters of the North and Baltic Seas will continue to see the rapid development of wind turbines fixed to the seafloor. Increasingly they will be connected to multiple countries, so the energy can be used wherever demand is greatest at any given moment. One of the first of these internationally connected windfarms is the 600MW Kriegers Falk windfarm in the Baltic Sea. It is now operating, and is connected to the grids of Denmark, Sweden and Germany. Kriegers

Falk is in Danish territorial waters, and the Danes are planning some much larger and internationally connected systems with some including the construction of artificial islands to act as hubs for grid connections, servicing and maintenance operations and also probably for hydrogen production. They are proposing to build an artificial island in the North Sea 80kms off the coast of Jutland, to act as a hub for 10GW of offshore wind farms. Cables will connect to Germany, Belgium, and probably other countries, and also back to the Danish mainland. Over in the Baltic the Danes are planning to use the island of Bornholm in a similar way as a wind power hub, in this case linked to Germany, Poland and Sweden.

In 2016 the Dutch grid operator Tennet proposed building an artificial island on the Dogger Bank in the North Sea to act as a hub linking Germany, Denmark, Holland, Norway and the UK. Back in 2009 the Danish architects Gottlieb Paludan proposed building large artificial islands as tidal lagoons to generate tidal electricity. None of these were built, but to me they all seemed excellent ideas, and could be developed with funding from my proposed Global Green New Deal. So a vast island with a tidal lagoon, probably also pumped storage hydro, hydrogen production and much else could be developed as a power hub linking to all the countries around the North Sea, and it would be additionally useful to connect to Iceland with its hydro and geothermal resource. I'll write more about this in Chapter Five as it could be used for many more functions than just energy production.

The Atlantic Ocean off the coast of Ireland and Scotland is the windiest place in Europe. The seas here are too deep for wind turbines to be fixed to the sea floor, as they are in most of the North Sea. In 2017 the world's first floating wind turbines started generating electricity off Peterhead, Scotland, in the North Sea.[41] In 2021 plans were implemented to massively ramp up floating wind with, for example, the planned 1.4 GW floating windfarm off the coast of Ireland, (and with this would come the replacement of the Moneypoint coal-fired power station) a green hydrogen hub, and the existing electricity grid moving from coal to wind-generated electricity.[42] Many countries wish to develop offshore wind but lack the suitable shallow seas such as the North Sea. For them floating wind turbines look very promising, and I would expect Japan, California, Hawaii and the Breton coasts of northwest France to be likely places for large-scale floating wind farms to be deployed first, but many other locations are also possible early adopters.

So far I have only discussed offshore wind. Onshore wind is of course just as important. In the UK context, where we have a dense population and high landscape value, I favour only a very slow and careful expansion of onshore wind power, while advocating a massive expansion of offshore windfarms. In many other locations, such as the plains of the American Midwest and the Mongolian Steppe, large-scale deployment of windfarms seems very sensible.

In Sweden they are starting to build wind turbine towers made from cross-laminated wood. These look like they will be lighter, cheaper and stronger and also entail fewer carbon emissions than steel. They may also be the basis for taller towers. Wind turbine blades may also get lighter, stronger and cheaper by being made out of new composite materials, such as those being pioneered by Scottish start-up ACT Blade.

Globally both offshore and onshore wind are growing fast, and this pace of deployment should be increased. In a later section, on patterns of ownership, I shall discuss how this could be done to achieve the best social and environmental benefits.

All this discussion has been about the power of the wind to make electricity. Of course there have been many older uses of the power of the wind, to propel sailing ships, drive wind pumps or to grind corn, to help dry our washing and dry foodstuffs. Many of these traditional uses of wind power may be due for a new wave of expansion, and I shall mention a few in subsequent parts of this book.

Hydro and Marine Power

Like wind, the power of water has had many traditional uses, with watermills the most obvious example. Since the late nineteenth century it has been used to generate electricity. It remains the largest source of renewable electricity in the world, and is still increasing, although less rapidly than wind and solar.

Hydro-electric dams on rivers are by far the biggest way in which water power is currently used to produce electricity. This can be done well, or very badly. The Balbina[43] dam in Brazil, on the Uatuma tributary of the Amazon, is an example of the very worst kind. Because the valley is wide and shallow the head is low, so it only produces 250MW of electricity yet it destroyed a large area of rainforest and displaced local people. The local climate having wet and dry seasons, coupled with the wide and shallow valley, means vegetation colonizes the reservoir bed as the waters recede in the dry season and then is flooded when the wet season arrives. This means large amounts of vegetation rot in the water, releasing so much methane that from a climate-change point of view this dam is unbelievably ten times worse than a coal-fired power station would have been.

The very best uses of hydro-electric power are in pumped storage systems, and Dinorwig[44] in North Wales and Kvilldal[45] in Norway are two of the best. Dinorwig has a reservoir deep inside a mountain in old slate workings, while Kvilldal is in a high and steep rocky valley in the mountains, so neither of them has the problem of rotting vegetation leading to methane release. Pumped storage means that in times of high electricity demand they release water down pipes and through turbines to generate electricity, then at times of oversupply from other sources such as wind, or at times of low demand such as the middle of the night, they use electricity to pump water back from a lower reservoir up to the top reservoir. This means they act like huge batteries storing energy. This is tremendously helpful, especially as wind and solar are both quite intermittent energy sources.

Most hydro-electric dams have both positive and negative factors. Most generate useful amounts of electricity without releasing significant amounts of greenhouse gases, but they do interrupt the ecological balance of the rivers, displace human settlements, and in some cases so drastically reduce the water flow in the lower river as to cause it to dry up, as has happened with the Colorado River in the American southwest. The Three Gorges Dam on the Yangtze River in China has helped reduce flooding and water shortages downstream, has improved ship navigation on the Yangtze and generates a lot of useful electricity. Most Chinese people look to it with pride. However the ecological and human disruption was considerable and it has one other unlikely, but potentially catastrophic, risk factor. If it collapsed, for example due to an earthquake, millions of people could be swept away in the resulting tsunami.

The vast majority of the world's hydro-electric dams are not currently used for pumped storage, but many could be converted into pumped storage systems as the need for energy storage increases as we build ever more wind and solar power. This conversion from ordinary hydro into pumped storage would be hugely beneficial. New pumped storage systems are being built, but suitable geographical terrain is currently a limiting factor. It is possible that by building underground storage reservoirs these geographical limitations could be overcome. Several organizations are planning to develop such systems, but, as far as I am aware, none has yet been built.

So-called 'run of river' hydro dispenses with the need for high dams and large reservoirs, but this limits the amount of power they generate and makes it much less reliable as it is dependent on the fluctuating volume of water in the river at any one time. As the cost of solar and wind continues to fall it seems unlikely that many more of this type of hydro will be built, but no doubt there will be some in particular locations where rivers are most suitable. It will not play a major role in terms of the global energy supply.

There are many other ways in which the power of water can be used to generate electricity, but none has yet been widely developed and deployed at sufficient scale to make any impact on global energy supply. However several technologies look promising.

The Rance tidal power station in Brittany, France, has a capacity of 240MW, and when it opened in 1966 it was the world's first tidal power station. A few more systems have been built since. By damming a tidal estuary electricity can be generated on the incoming and outgoing tides, but at great cost to the river's ecology. There have been numerous plans to build a dam across the Severn Estuary in Britain. This would be like the Rance scheme but on a very much larger scale. There are other ways to use the power of tides to generate electricity without damming rivers and disrupting their ecology. Some potential projects even have the possibility of many ecological and economic benefits, as well as generating useful amounts of electricity.

The Swansea Bay tidal lagoon[46] is a project that has planning permission, but has not yet been built. The idea is to build a U-shaped barrage, with both ends joined to the coast and looping out to sea. This would enclose a tidal lagoon, generating electricity on the incoming and outgoing tides. The lagoon could have many uses, as a site for sailing and other leisure activities, or as a site for aquaculture, or to construct enhanced habitats for wildlife, or for some combination of all of these. This is a project that should go ahead with all due speed, and if it proves as useful and beneficial as I and many of its supporters think possible, then it should be the first of many. The people behind this project have identified half a dozen potential tidal lagoon sites in the UK and more in other countries. In the section on wind power I mentioned plans by the Danes and the Dutch to build artificial islands in the North Sea to act as hubs for wind power and hydrogen production, and speculated that they could incorporate tidal energy and pumped storage, as proposed some years ago by the Danish architects Gottlieb Paludan. I will elaborate more on these possibilities in Chapter Five.

There are all sorts of other ways in which water can be used to generate electricity. Tidal stream turbines can be deployed where there are naturally strong tidal currents, and this has successfully been done in the Pentland Firth between Orkney and the Scottish mainland. There are a number of wave power devices being developed and trialled. In the Orkney Islands is the European Centre for Marine Energy[47] which is at the heart of all this tidal and wave research and development. It is precisely the kind of project my envisaged Global Green New Deal would see massively scaled up and replicated, with projects fast tracked into full-scale deployment.

It is not just in electricity generation that the power of water can be used. Marine and river-source heat pumps can be used to heat individual buildings or, via district heating systems, whole town and cities. The city of Drammen in Norway gets most of the heat it needs from the local fjord by this method.[48] I see this kind of technology having huge potential in many locations and it is just the sort of technology that could do with a massive boost of investment.

Geothermal

Generating electricity from geothermal power provides a significant portion of the electricity supply in only about half a dozen countries: Iceland, Philippines, El Salvador, Costa Rica, Kenya and New Zealand. The USA generates the most geothermal energy, but due to its vast energy demand, geothermal as a share of overall electricity supply is only 0.3%. Due to the falling costs of solar and wind it seems unlikely that geothermal energy will play a huge role in the global electricity supply, but changes in technology and costs could well change that.

Some people talk of ground-source heat pumps as a form of geothermal energy, but this is very different technology to deep geothermal energy, and they are used in very different ways. I will return to heat pumps in the section on the built environment. They probably have a huge role to play, not in electricity production but in heating buildings, sometimes as part of new district heating systems.

Biomass, Bioenergy, Biofuels and Biogas

Bioenergy is a term that covers an extraordinary range of different technologies, all of which convert biomass to energy. This can be for heat, electricity or transport fuels and it can refer to something as low tech as a campfire to a great range of sophisticated technologies seeking to utilize as much of the potential of the biomass to simultaneously create an optimum mix of electricity, heat and transport fuels. Biomass is living, or recently living, biological matter, overwhelmingly from plants, and is therefore a renewable form of energy. However there is much debate about which types of bioenergy are really ecologically sustainable. Certain technological developments in the bioenergy field seem to me to be of tremendous potential benefit while others represent a further onslaught on the natural world. Unfortunately most current development is happening in the least beneficial areas. Let us flag up four examples of what we are currently doing which would be better not done at all.

Firstly Indonesian rainforest is being cleared to produce palm oil as a transport fuel for people in the rich countries of the world. This ecologically destructive practice makes no sense and it even fails to produce truly low carbon fuels. Better to leave the forests intact and to develop new forms of diverse and sustainable gardening and cropping within these bountiful ecosystems.

The second area of reckless and unsustainable development of biofuels has been using annual human food crops such as maize, wheat, sugar cane and soya beans to make transport fuels. The inflated energy demands of rich motorists has pushed up the price of basic food stuffs, which in a world with a billion people already struggling to feed themselves seems particularly immoral. Also the processes involved do not make much sense in energy terms as a lot of fossil fuel and water are consumed in the process of growing these crops for biofuels.

The third area of concern is how anaerobic digesters have been used to produce methane. Too often maize and other annual crops, requiring heavy use of chemicals, ploughing and other ecologically unsound techniques (and of course also competing with food production for fertile land) have been the preferred feedstock. Better to have used food waste, agricultural waste or sewage, or to have used perennial crops like grass.

The fourth area of poor use of bioenergy is the combustion of woodchip to generate electricity as has been done on a vast scale at the Drax power station in the UK. Ecologically rich forests in USA have been clear-felled and transported half way across the planet just to be burnt to generate electricity.

These examples of poor use of bioenergy have caused many people to oppose all use of biomass to create energy. However, some uses of anaerobic digesters, and even of combustion of woodchip, are benificial, but they do need to be carefully considered. Extracting biomethane from sewage works and landfill sites and feeding this into the gas mains seems sensible and has been quite widely developed. There are some uses of biomass that could be greatly expanded with many potential benefits.

Algal bioreactors are one of those technologies that seems to have much better potential, as algae grow about thirty times faster than any of the more widely used crops. Various forms of seaweed or micro-algae have been used directly as human food, or as the basis for animal feeds, transport fuels, cosmetics and to make compost, and also to remove carbon dioxide from factory chimneys. As algae can be grown in seawater, or polluted water, in deserts or on factory roofs, they do not have to compete with food crops for fertile land, nor do they necessitate the destruction of ecologically rich habitats as has been the case with oil palm or woodchip.

Another use of biomass that has huge potential is for producing biochar. This is essentially charcoal made to be used in agriculture to improve soil structure, fertility and soil-based carbon sequestration. When wood is burnt in an oxygen deprived process it turns to charcoal, and it may be possible to do this in a way where the heat could be captured and used in a district heating network or also to make electricity, but these would be almost by-products from the biochar production, rather than the main purpose of using the wood or other biomass.

Under my envisaged Global Green New Deal the existing and wide scale poor uses of biomass such as the four examples I have cited would be massively scaled back or totally eliminated. Funding would be focused on research and development of biochar and algae. We will look again at these two technologies in the following chapter on food and farming as their potential uses are greater in these sectors than as major energy sources, although in the future algal bioreactors may make a useful contribution here too.

In many poorer parts of the world where people lack electricity, various forms of biomass are used for cooking, and to some extent also for heating and lighting. Gathering wood to be used directly as fuel, or made into charcoal, is having devastating effects as precious trees are felled, increasing desertification. In some areas this is a huge drain on people's time and effort. Often manure is used as a fuel when it would be better composted and used in vegetable gardens. Millions of people die each year due to air pollution, and in communities where most people cook on biomass this is the main factor in so much respiratory illness and death. Rural electrification is happening anyway, but would be greatly speeded up under the envisaged Global Green New Deal as solar power and agrivoltaics are rapidly deployed, especially in those poor parts of the world where people currently cook with wood, charcoal or dung.

The combustion of wood, be it in wildfires, controlled use of fire such as on grouse moors, or burnt on domestic stoves and also in woodchip power stations (if not fitted with the best smoke-scrubbing technology) produces black carbon, which contributes to global warming by reducing the albedo effect. We need to stop, or at least massively reduce our burning of stuff, be it fossil fuels or biomass. There are better ways to make useful energy, be it for electricity, heating or transport, and better ways to manage our land.

Peat, Brown and Black Coal, Oil, Gas, Nuclear

All these sources of energy will be phased out as we shift into 'The Solar Age', but at varying speeds of urgency. Peat only plays a relatively minor role in energy generation, and mainly in a few countries, notably Ireland and Finland. It should be halted immediately. Intact wet peat bogs are vital carbon sinks and rich ecological habitats. Like selling peat in garden centres, burning it as fuel should have stopped decades ago.

For many countries coal has been the main source of electricity for a very long time. All countries need to phase it out, and progress on this has been a very mixed picture. Belgium, Austria, Sweden and Portugal have all closed their last coal-fired power stations. Over the coming few years many more countries will. Coal use is only really still significant in four of the EU countries; Germany, Poland, the Czech Republic and

Bulgaria. The new government of Germany, encouraged by the Greens within the coalition, have brought forward the planned coal phase-out date from 2038 to 2030. This is an important step in the right direction, but I would like to see coal phase-out across the whole EU before that, perhaps by 2026. This would be challenging, but possible.

My proposed Global Green New Deal would seek to speed up a global phase-out of coal as quickly as possible, and given the right political leadership and investments, this could be done very much more quickly than almost any politicians or commentators are suggesting. For example Australia and South Africa are two heavily coal dependent economies that each has excellent solar resources. If they underwent the speed of economic restructuring that the USA achieved in the six months or so after Pearl Harbour they might be able to quit coal over the space of a year or so. They could achieve many goals including increased food production and water security, cleaner energy and many new jobs and economic opportunities as they co-develop solar power and agriculture in agrivoltaic systems.

The phasing out of oil and gas use will take a bit longer than coal or peat, but it could be done very much more quickly than most people expect. Oil is mainly used as fuel for transport, and I will look at its phase-out in the section on transport. Oil is still used to generate electricity, especially in small generators in off-grid situations. Nearly everywhere it makes sense to switch to renewables plus storage immediately. Gas is mainly used for heating, and that is a topic we will cover in the section on the built environment. However, gas is still a major source of generating electricity. As I write, gas prices have skyrocketed and countries that already made the switch from gas to renewables are profiting handsomely. The falling price of renewables means that even if gas prices fall back to where they were it will still not make sense to keep using it, and the kinds of proposal in my envisaged Global Green New Deal would ensure it was left in the ground. Phasing out coal use is the first step, but phasing out gas and oil need to follow as quickly as possible.

Nuclear power still has its advocates. Clean reliable nuclear energy has been an idea that has been hard sold to us for decades. Nuclear fusion has been the promising technology that is only a few years away ever since I was in school in the 1960s. It remains a distant prospect as a practical source of energy, despite many billions of pounds of investment. The decommissioning of Sellafield is predicted to cost at least one hundred billion pounds and take one hundred years, and many of the radioactive isotopes will remain hazardous for hundreds of thousands of years. Creating onerous responsibilities and costs for generations to come seems to me utterly irresponsible. 'Atoms for peace' was always a front for weapons production. The whole sector has had massive subsidies, and the costs of cleaning up the mess are being left to the public.

Hinkley C is the huge nuclear power station currently under construction in Somerset. It is an extraordinarily expensive mistake. The cost of generating electricity with offshore wind is already considerably cheaper than the agreed price for the as yet to be generated nuclear electricity from Hinkley, and as the price of renewable energy continues to fall, by the 2030s the cost difference will be immense. The UK government will be tied into contracts to buy electricity at probably over ten times the market price.

Some advocates talk with enthusiasm of small modular nuclear reactors currently being developed by Rolls Royce. I remain sceptical. I have heard so much spin from governments and public relations industries trying to sell us nuclear power for about seventy years now. The risks remain as potent as ever and the benefits as illusory as ever. Given the enormous and rapid advances in renewable energy, energy storage, grid connection and a great range of other related cleantech, nuclear looks increasingly costly and irrelevant.

The underlying assumption behind my envisaged Global Green New Deal is that no new investments are made in coal, oil, and gas or nuclear. Instead, massively increased effort and investment should be made in reigning in waste and keeping global energy use from growing too much, which will mean falling use per capita in the rich world in order to allow the rapid spread of electricity to the billion or so who currently do not have access. Simultaneously, of course, massive investments should be made ramping up renewables. Investing many trillions of pounds, dollars and Euros in renewable forms of energy opens up the prospect of experimenting with radically new concepts of ownership.

Co-operative Models

In the past coal, oil and gas have been extracted from the Earth and the profits went entirely to the companies involved, with usually the host nations taking a cut in taxes and licence fees, but also considerable flows of public money used to subsidize developments. Currently renewables are usually developed along similar lines, with minimal support from or benefit to local communities. Where community ownership of renewable energy has been established, people benefit in numerous ways. In Fintry, Stirlingshire, Scotland, a commercial wind farm was to be built on the nearby hills and instead of opposing this development the local community bought one of the turbines and formed the Fintry Development Trust.[49] Income from sales of electricity is transforming the community; they are super-insulating their houses, investing in domestic solar water and photovoltaic panels and have had several ambitious plans for all sorts of exciting projects such as producing their own wind derived hydrogen, starting their own hydrogen powered bus service to Glasgow and building a district heating system in the village. I love this degree of community ambition.

Community ownership of renewable energy projects in the UK is still relatively rare, and government policy has often made setting up these kinds of projects very much more difficult than it ought to be. I am a member of half a dozen renewable energy coops. They tend to be rather small scale, and so their financial surplus and community ambition is necessarily limited. Also because they are so small scale they rely on volunteers to do the administration of the projects, which is sometimes rather onerous. The solar cooperatives that I am a member of range between 49kW and 91kW installed capacity. Working at a larger scale has many advantages.

Awel Aman Tawe is an amazing Welsh charitable organization that does great educational work around climate change and has initiated some excellent projects. It set up the Awel as an energy co-op to build and run two Enercon 2.35MW wind turbines at Mynedd y Gwrhyd, near their headquarters at Cwmllynfell, twenty miles north of Swansea in South Wales. They have also set up Egni, the UK's largest rooftop solar co-op, with 88 photovoltaic systems on schools, village halls and other community buildings across South Wales, including the huge Newport velodrome. These 88 solar roofs together have an installed capacity of 4.4MW. Egni's latest share offer is on course to raise another £4.8 million. Operating at this scale has many advantages over the smaller separate co-ops that I am a member of. It allows them to have paid staff, to work on a wide range of charitable work and so to access charitable and education funding streams. Being based in Wales, and deeply rooted in their community helps. The Welsh government is very much more supportive than the UK government would be of any such project starting up on the English side of the border.

The Big Solar Co-op is pioneering a new way to set up and run a larger scale solar energy co-op engaging with a huge number of solar enthusiasts and volunteers across the country to get solar roofs installed all across the UK. They initially just have a couple of paid coordinators, and the project is still at an early stage, so they do not as yet have any solar roofs up and running, but it does have the potential to really take off. When Wikipedia first started the idea that anyone could create an online encyclopaedia put together largely by a vast global network of volunteers was laughed at and dismissed as impossible and a utopian fantasy. It has out-competed all conventional encyclopaedias written by paid experts. Could a global energy system be organized with this kind of structure? We won't know unless we try to do it.

Across Europe there are countless examples of communities that have experienced multiple benefits from community ownership of renewable energy projects, many on a bigger scale, and most with very much more supportive government policies in place. There are quite a number of different models for community ownership, from the small groups in villages acting on their own initiative up to larger municipal authority led projects. Many of the best examples are in Denmark, where local parish councils have the option to have some degree of community ownership and benefit on any renewable energy project larger than individual rooftop solar panels. That is a model I would seek to take on to the global stage.

Middelgrunden wind farm is a 40MW offshore windfarm that opened in the year 2000, just off the coast from Copenhagen. It is 50% owned by a co-operative of ten thousand mainly local investors and 50% by the local municipally owned utility company. In Germany, the Trianel Windpark Borkum takes municipal ownership up to a larger scale. The first phase was opened in 2015 and the second in 2021, each of 200MW. It is located to the north of the island of Borkum on Germany's North Sea coast. Dozens of local municipal utilities from across Germany, and a number from surrounding countries, have a stake in this 400MW windfarm. Trianel as an organisation exists to assist the co-operation of all these municipal authorities.

In Germany local Stadtwerke are companies owned communally by the people of a given area, sometimes combining private sector investment with majority local authority ownership and control. Traditionally they are responsible for all sorts of local services from public transport to sewage, and energy generation and distribution within a local geographically defined area, usually a single town or city. In some places they also run hospitals, care homes, manage much of the local housing and may even own farms. In Britain we have largely forgotten what a powerful municipal sector we had from the 1870s until the 1940s. As mayor of Birmingham, Joseph Chamberlain led an extraordinary expansion in the provision of gas mains, sewers, clean water, slum clearance, improved education, the founding of the University of Birmingham and much else. Chamberlain stated that 'We have not the slightest intention of making profit... We shall get our profit indirectly in the comfort of the town and in the health of the inhabitants' [50] Can we imagine what a global energy economy, infrastructure and service provision would look like if it were designed for the comfort and health of all of humanity?

It is my proposal that humanity makes the transition to an economy based on 100% renewable sources of energy, and that all of this has some direct community benefit and ownership. Exactly how this happens will vary with the scale, type and location of the project. Let us imagine that our global network of 39,000 communities each had a municipal sector with the kind of vision and competence evidenced by Joseph Chamberlain's time as mayor of Birmingham. Wealth, carbon and the other taxes we have discussed would be raised globally and invested via these 39,000 community owned organizations based on the German Stadtwerke, or the kinds of municipal structures that flourished in Birmingham in the late nineteenth and early twentieth centuries. Our global network of communities would be linked together via some kind of loose network like that established under the Aalborg Commitments. Our global network of communities would be continually seeking to learn from each other. Educational exchange between them would see many young people developing their careers spending years studying in universities and working on projects in other communities all over the world, learning and gaining knowledge of best practice across a wide range of fields that they could then take back into their own communities.

In many of the poorer parts of the world these municipal investments would be focused on making sure that everybody had good access to clean water, good food, sanitation, electricity, health and education. The poorest billion people would see their lives dramatically improved. Our focus in this chapter is on energy. Most rural communities could easily supply themselves with local renewable energy, and in the case of much of the world this would be best done predominantly with agrivoltaics.

Big urban centres would need to import energy from further afield, as would whole countries that are densely populated and heavily industrialized with relatively poor wind and solar resources. Germany is the most obvious example, but Japan, the eastern seaboard of China and the urban area that extends from Washington to Boston in the north-eastern USA would all fit into this category of big energy importers. As we have noted some areas of the world will inevitably become big exporters of renewable energy, either exporting electricity via HVDC cable, or via hydrogen along pipelines or in the form of ammonia or methanol, which are easier to contain on tankers.

So far, as far as I am aware, the 400MW Trianel windfarm just north of the island of Borkum in the North Sea is the world's largest renewable energy project that is owned by a network of local municipal organizations. I want us now to envisage a few hundred projects each perhaps ten times as big, so about 4GW each. Some might be floating wind projects stretching in an arc from the coast of Brittany, up the Atlantic, off the coasts Ireland and Scotland to Shetland and on into the North Sea. Windfarms in the North Sea and Baltic Sea would also be added, but here, with shallower seas, the turbines could be fixed to the seafloor. Norwegian hydro and Icelandic hydro and geothermal would add to the mix. Solar energy from the south, from Morocco in the west, through North Africa and the Middle East would all be linked in to this network, supplying local energy needs and exporting energy to the power hungry cities of Europe.

Australian solar, offshore wind from the East China Sea and the Pacific, tidal energy from Indonesia and geothermal energy from the Philippines and wind from the Mongolian steppes would all help provide power to the energy hungry cities of Japan, Korea and China. Offshore wind would have a major role to play in powering the big cities on both coasts of the USA.

Can we imagine a global renewable energy co-operative? It would have as shareholders all of humanity; all 7.8 billion of us would have an equal sized shareholding given to us, and from which each of us would receive a small income. This might be in the form of Universal Basic Income, or in addition to it. This global cooperative might get huge initial funding via global taxes on wealth and on resource use, but as it develops the main income would come from energy sales.

According to the International Renewable Energy Agency as of the end of 2020 there was 2,799GW of installed renewable energy capacity in the world. [51] 43% of this was from hydro, and 26% each from solar and wind, with 5% coming from a mix of other sources. If our global co-op set up seven hundred of these 4GW projects, that would mean a doubling of installed capacity for renewable energy. Most of this would be solar, with quite a lot of wind and some tidal, geothermal and other technologies. Additionally there would be very many, perhaps millions, of smaller projects located in communities and serving their local needs. Ideally many of these would embody the spirit and ambition of tiny projects like Fintry and the somewhat larger Awel Aman Tawe, but with very much more help and assistance from the network of similar projects and generous funding from globally redistributive taxation and supportive legal and policy frameworks.

Transport

While decarbonisation had progressed well in the electricity sector over the last decade, no similar progress has been achieved in the transport sector. No country has yet seen significant and sustained falling emissions in its transport sector. There are many local good news stories of cities cutting traffic, of technical innovation and of behavioural change, but so far no massive cuts in transport related emissions. That needs to change, and to do so progress needs to be made on many fronts.

Much can be done with technological change, but yet more can be achieved through behavioural change. Of course we need both, and the relationship between these two is complex. Many of our personal choices are determined by what is available in terms of technology, price and accessibility, and those are all determined to a large extent by political choices, hence the need for the Global Green New Deal.

A Note on Weight, Speed, Time and Fun

If we learn a few simple lessons from physics and apply them to our transport planning and technology development we could cut energy use significantly. When moving anything, be it a car, truck, train, ship or aeroplane, much energy is used to displace the air or water it is passing through, and in friction with the surface it is passing over. Travelling more slowly will always use less energy than travelling fast, and lighter vehicles and vessels will always use less energy than heavier ones. Steel wheels on steel rails will always be more efficient than rubber tyres on tarmac as the friction is much less. Aerodynamics and shape also matter, but it is weight and speed where we really need to rethink things.

Many trends in recent years have seen the development of the most extravagantly wasteful types of transport. The fashion for big SUVs instead of small cars has many negative impacts, as their increased weight makes them less energy efficient and also more damaging in the event of accidents. Concorde was much less energy efficient than other planes that travelled more slowly. High speed rail, such as the controversial HS2, will use more energy than slower trains. Ever increasing the speed at which people travel has been a key theme of technical development and transport infrastructure planning ever since the industrial revolution. The justification for this has been that it saves time. We should now reverse this logic. Let us save more time by eliminating unnecessary and often boring journeys. Let us plan new methods of transport that are slower, lighter in weight, more energy efficient, less polluting, safer and more fun.

Bringing concepts of fun, conviviality, health, fitness and wellbeing into transport planning is proving revolutionary. Many cities Dutch cities such as Amsterdam, Utrecht and Groningen, and other cities such as Oslo, Copenhagen and Barcelona, have become very much more pleasant places to live. They have done this by reducing car use and making walking, cycling and public transport of the highest standard. Car parks have been changed to green spaces, with plenty of public seating, nice cafes and interesting cultural activity and more spaces for children to safely play. Paris is just starting to seriously make this model shift and follow the Dutch model. All cities should follow suit. The benefits are many.

In this era of extreme waste of energy, flying off for a fortnight, a week, or even a weekend in some far off land has become normal. What if we reimagine the concept of fun and of travel? Might the norm for a family holiday in future be walking the slow ways network[52] directly from their home to see some relatives in the next town, or cycling Sustrans routes to see more distant friends, or explore new landscapes. Foreign travel can have many educational benefits, so why not take a year out and go in a sailing ship, or a hydrogen fuel cell ship, to Africa or India and participate in an interesting and worthwhile project? One of the key concepts in my Global Green New Deal is to reimagine how we utilize the human resource: what if all 7.8 billion of us had equal opportunities for paid sabbatical years, time we could devote to travelling slowly and learning much.

Road Transport

In June 2020 there were 38.4 million licensed vehicles in the UK, according to the RAC. The average car driver clocks up about 10,000 miles per year. Most cars are parked for about 97% of the time. Of the 3% of the time that they are used they mainly do repetitive journeys like commuting, or the school run. To cut carbon emissions and local air pollution, to free up valuable urban space and to improve the quality of life in our cities we need to massively reduce the number of vehicles. The planning process will be vital to create the fifteen-minute city. Any new housing must have work places within walking distance, good provision of cycle ways and bus, tram or train routes. Localizing the economy and integrating work places and housing could massively reduce the need to commute, and making safe and enjoyable routes to schools should entirely eliminate the school run. As the amount of time most people spend in cars declines, the advantages of shifting from individual car ownership to car sharing becomes ever greater. Time spent commuting to work is time wasted, and cutting commuting distances is just as important as modal shift. I explore these issues in more depth in Chapter Five, looking particularly at how the Global Green New Deal might impact Herefordshire.

Having said all that about the importance of the need for car and truck use to decline, it is still important to stress the aspect of technological change. Norway is leading the world in the transition from petrol and diesel to battery electric cars, and for those cars to be using renewably generated electricity. Other countries will follow as battery technology improves and the pressure for decreased pollution grows. However electric cars still have carbon emissions embodied in their manufacture, and particulate pollution from their tyres and brakes. Heavier vehicles are worse in all these regards, so a Tesla has a much bigger ecological footprint than the new very lightweight and aerodynamic Aptere. The Aptere claims to have a thousand mile range, and as it has solar modules built into the body of the car it can re-charge itself just by being parked in the sun. In Germany Sono motors have developed another car with integrated solar cells. This is likely to be a major trend in future.

Back in the 1960s and 70s electric milk floats were very common. They had heavy lead acid batteries. The growth in battery electric cars over the last decade has been due to the falling cost and improving performance of lithium ion batteries, which carry more power relative to their weight. There are many other types of battery, and flow batteries seem to have several advantages over lithium ion batteries. In flow batteries electrical energy is stored in a liquid electrolyte in simple storage tanks which makes cheap, large scale, long term electrical storage possible in a way that is unlikely ever to be possible with lithium ion batteries. This will probably be developed for grid scale electricity storage over the next few years. It may also become important in the transport sector, as the electrolyte can be changed more quickly than charging an ion lithium battery.

Hydrogen fuel cells are very much less well known and understood than batteries, but like batteries they are a vital part of the 100% renewables vision, for energy storage and many other uses including many forms of transport. The Riversimple Rasa is probably the most sustainable car yet designed for a multitude of reasons. Riversimple are adopting a circular model, leasing cars rather than selling them, keeping the responsibility for repairs and maintenance in house. Companies selling cars have always had economics pushing them toward built-in obsolescence, whereas the economic model adopted by Riversimple is designed to maximize durability and reliability. JCB calculated that for some of their heavy vehicles to go electric they would require a four tonne lithium battery, which is simply impractical. They looked at moving to hydrogen fuel cell but have opted for the cheaper alternative, burning hydrogen in an internal combustion engine in much the same way as their existing diesel vehicles do. There is a race on between hydrogen and batteries for the mass markets of cars, buses and trucks, but both systems have advantages, depending on the type of vehicle and how it is used. As the prices are falling and the technology improving in both sectors it is too early to say which will become more important.

International Travel

Air travel is of course particularly damaging. Although some technical progress has been made with very small planes flying very short routes changing to battery electric, the chances of any form of ecologically sustainable mass air transport for distances from the UK to the Mediterranean, for example, is very unlikely in the short to medium term. We need to reimagine how we spend our holidays. Instead of weekend stag does in Prague or a fortnight family holiday in Spain we might travel much less distance, yet have a much more interesting time. One of the most exciting transport developments of 2020 has been the emergence of the Slow Ways network. Led by Dan Raven-Ellison, thousands of volunteers have been plotting new uses for our footpath network, using digital tools to share information, linking up towns and cities and promoting exciting journeys of exploration. Many people in the UK hardly know their own region and might enjoy discovering its complexities and richness.

Short business trips make up a significant share of flights. Zoom and other online tools could replace many of these trips, and relocalizing a greater share of the economy would also reduce the need to fly. Increased taxation would also suppress demand. At some point the total ban on all fossil fuel air transport will probably be necessary, or at least some form of rationing, and as a first step a frequent flyer levy should be introduced immediately. Cleaner technology for long-haul flights may eventually be achieved but this is probably decades away, and addressing the climate crisis simply cannot wait that long. Many people are working on cleaner ways to fly, and they may make more rapid progress than currently seems feasible to me, in which case I'd happily be proved wrong.

Travelling to other countries can have tremendous educational benefits, and trade too can be highly beneficial. We need to do it differently. Flying to New York for the weekend is never going to be sustainable, but spending a year living and working in the USA may have many benefits. Travelling slowly can have many advantages to travelling quickly. We should make fewer journeys, but make sure that those journeys are deeply worthwhile.

Bill McKibben points out that transporting coal, oil, gas and now a bit of woodchip makes up very nearly 40% of global tonnage of shipping.[53] As we switch from fossil fuels to renewables we may need a lot less shipping. Then if we localize a lot more production we might further decrease the number of ships. Currently shipping uses a very dirty form of bunker fuel, releasing a lot of pollutants including sulphur dioxide, nitrogen oxide, carbon monoxide, carbon dioxide and particulates. Diesel powered ships also create noise pollution, oil spills and collisions with wildlife. We could do better.

Ocean going ships can and will be designed to be very much less polluting long before any such technological innovation is possible in the field of air transport. Electric batteries are already powering many ferries in Scandinavia for shortish journeys. Maersk, the world's biggest shipping company, has ordered six large container ships with methanol fuel cell propulsion. The methanol will be made from hydrogen derived from solar and wind power via electrolysis, and mixed with carbon dioxide captured from industrial processes to make the methanol. Following on from Maersk's announcement came a similar one from X-Press Feeders for eight container ships utilizing methanol. Could all long distance shipping go this route toward green methanol fuel cell propulsion, possibly with wind assistance in the form of sails, as some are advocating? This would reduce emissions of everything that we listed as being so bad with diesel powered shipping.

If we are all to have the occasional sabbatical year off to travel and study might the journeys be by passenger ferries and trains rather than planes? It would be much quicker and easier to transform the world's shipping and railways than to make air travel sustainable. I look forward to seeing ships combining some cheap long distance passengers with cargo.

The Built Environment

The scope for improvement to our global housing stock, and the design and construction of our cities, is almost infinite. In our current market driven economy vast amounts of money are flowing into providing more and more luxurious ways of living for the wealthy, while the poorest people in the world and indeed in many of the richest countries are neglected. My proposed Global Green New Deal would reverse this by investing on a more or less equal basis for everyone on Earth. A lot of the funding would be focused on the public realm to improve the kinds of infrastructure that better quality of life is dependent upon, from sewers to tram systems, from libraries to parks, sports facilities to care farms and of course schools, universities and hospitals.

In colder climates improving the thermal efficiency of our housing stock is of critical importance. Individual houses, schools and other large buildings, and indeed whole towns and cities, can now be designed and built to use very much less energy than they used to do. Building standards legislation in many countries is slowly raising the legal minimum, and pioneering architects and builders are pushing the boundaries of better building. In passive house design the idea is to combine super insulation and air tightness with optimising passive solar gain by having south facing windows and so make buildings that are warm and comfortable without using much energy.

Many new buildings seek to build upon this increased energy efficiency by adding solar panels, and in an energy positive building this on-site generation produces more than the building uses. I have been in a few buildings which receive energy payments rather than bills as they generate more energy than they use. In many cases as well as generating their own electricity and heating they also are recharging an electric car, and still generating surplus electricity for much of the year. Most of the early pioneers were self builders. Now some of the best architectural practices and small innovative building companies are doing this, but so far none of the UK bulk house builders. It is technically possible and should rapidly become a global norm. The falling price and easy availability of solar panels means they should be built as standard in any new home just as much a functioning water supply or a proper toilet.

So let us imagine that all new housing built globally was designed to generate as much energy as it used for electricity, heating and also for recharging electric bikes and the occasional residual car use. To achieve this it is really important to think from early in the planning process. Almost all new buildings should be laid out to optimize roof orientation and so maximize the uses of solar energy for passive solar space heating and for solar photovoltaic panels, and if those solar panels are like Naked Energy's hybrid panels both hot water and electricity will be efficiently produced. In new build rural and suburban settings, achieving energy positive buildings is not that difficult.

In high rise city centres where space is limited, and in older buildings, achieving energy positive standards is not possible, or even desirable. Such areas and individual buildings will bring in electricity via the grid, and large scale district heating and cooling systems will also have a vital role to play. Improving the thermal efficiency, quality and comfort of our existing buildings, towns and cities will be a challenge.

Retrofitting individual houses in a street on a one by one basis is very expensive. A much better approach is to do whole streets, or indeed whole cities, as a single project. I would love to see a trial project on one of our old industrial cities and see what could be done in terms of this whole city approach. For example, Hull might chose to install a district heating and cooling system connected to every property in the city utilizing a large scale marine source heat pump to extract heat from the Humber. The Norwegian city of Drammen extracts 85% of its heat from the local fjord[54]. Hull could well do something similar, utilizing cheap electricity when North Sea wind farms are generating more than is needed. This surplus wind power can be converted to heat via a heat pump and stored as hot water until it is needed. At the same time as installing the whole city heat main, many other changes could be made: insulating and retrofitting the existing housing stock, improving public transport and active travel facilities, and creating new parks and all the other aspects of public infrastructure that make city living more pleasant.

In the section on transport I mentioned how many cities are following the Dutch model of reducing car use in city centres. Paris, under Anne Hidalgo's leadership, is now implementing the concept of the fifteen-minute city, where work, shops, schools and places of recreation and culture are all within a fifteen minute walk or cycle of where people live. This means a radical redistribution of services, serious investment in public transport and cycle paths, and in reducing the space given over to cars. It has involved removing tens of thousands of car parking spaces, the pedestrianization of many streets, slower speed limits and increases in the prices charged at the dwindling number of car parking meters. New parks are being created on old car parks and pedestrianized roads. Restricting car use near schools while making cycling proficiency lessons a normal part of primary education is important to reduce or eliminate the car-based school run. The city is becoming a cleaner, less polluted and a less stressful place to live. Over time this should bring many health benefits.

Many other cities are following suit, each in slightly different ways, and learning from each other as they evolve. There are some good examples of this trend on most continents. The scope to do more and more quickly is massive, and the changes envisaged in the Global Green New Deal seek to set in place the right macroeconomic levers to help bring this about. Local city-level political leadership is always of critical importance. The planning of what buildings we build and where, especially in our cities, should be informed by this concept of the fifteen minute city. Major investments should be made in the public sphere: in parks, libraries, art galleries, convivial places for people to meet, in public infrastructure such was water mains, sewers, district heating and cooling systems, cycle paths, trams, light rail and other public transport systems. Cities should become very much more pleasant places to live, with major gains in terms of reduced pollution and carbon emissions.

There is a plethora of terms describing various goals to make buildings more energy efficient, to use water and materials in a more sustainable way, to cut the carbon emissions generated in the materials from which the building is constructed, to use recycled materials and to improve the experience of living and working in them. Places like BedZed in Sutton, south London, which opened in 2002 was one of the early pioneers of One Planet Living. Biophilic design seeks to learn from nature and to bring plants into the heart of the built environment. Sustainable architecture seeks to do no harm while regenerative architecture, like regenerative farming, seeks to build more enduring benefits. The terminology keeps changing. Architecture in a global market economy follows the money, so more and more effort is put into minute details of housing provision for the rich while the poor are neglected. Better architecture for the masses is of course utterly dependent upon a transformation in global economic justice.

Everybody on earth should have access to pure water, safe sanitation and a reliable electricity supply. Currently 46% of the global population lack access safe sanitation,[55] 25% lack access to clean drinking water[56] and 13% lack access to electricity.[57] Providing these basic services to all of humanity should be a top priority, and it is not rocket science. Technically, universal provision of all three could easily be achieved in a very short time, perhaps just three, four or five years. Why this has not already been achieved is down to a lack of peace, good governance and economic justice. Once all these factors are in place building better housing, providing excellent health and education services, and the kinds of opportunities that most people want, ought not to be that difficult. I shall explore what might be possible in various sample locations in Chapter Five.

The recent history of how the British water and sewage systems have been mismanaged provides useful lessons in what not to do, and a hint of how things could have been done so much better. Privatization has led to massive under investment in new infrastructure. Many of the private water companies are based in offshore tax havens, or are part of global corporations which have little interest in the quality of local rivers or in the quality of our drinking water. Quarterly profits and paying shareholder dividends have been prioritized. The UK Environment Agency has had funding cuts so big it simply cannot do the tasks of water quality monitoring and enforcement of laws to limit pollution. Coupled with this has been the continued process of building more housing, concreting and tarmacking more surfaces and emptying storm drains into the sewage system instead of being separately dealt with.

The release of untreated sewage in our rivers and on our coasts has increased dramatically over the last few years. This needs to be reversed immediately. The ideas I am proposing will massively enhance local government and its role in raising standards, linking with other local governments that are already doing a much better job, for example in parts of Scandinavia. Huge funding would flow from the global level into our network of 39,000 local governments, many of which would of course be linked together as the 32 London Boroughs already are. Local government will invest heavily in better infrastructure globally, so no matter whether one lived in rural Africa or urban London one would have safe clean water, and sewage would be dealt with very much better than is now the case. Perhaps we would think differently about sewage. Instead considering it a waste product, circular economy thinking would prevail and it would be thought of as a resource, fed into anaerobic digesters to extract the methane, the residue then being composted and used in agriculture. Industrial wastes and storm water run-off would each need separate systems. In both cases, to some extent, they could become resources to be processed and used rather than just becoming waste products flowing into our sewage systems.

Throughout this book I argue for social and economic justice and for a rebalancing of the global economy. The private sector has a vital role to play, as does the public, and a greater emphasis on municipal and global organisations would be useful. Organizations responsible for monitoring pollution and enforcing standards need to be massively beefed up and expanded on to the global scale. The involvement of the private sector in water and sewage has been a disaster, as it has in the health and education sectors. In building and architecture the self employed and companies of all sizes have usually been at the forefront of good design and construction, but they need the right policy and planning guidelines, and these come from the various layers of government. Owning one's own house is a fairly common goal in life, and it ought to be possible for everyone on Earth. In different climatic and cultural conditions the building styles and objectives may be somewhat different, but everyone has the same basic requirements for shelter. Housing should be about providing people with homes, rather than as assets bought and sold primarily as a way of making money.

The changes to employment patterns envisaged in this book would see the loss of hundreds of millions of jobs in many sectors, from fossil fuels to banking and retail. Many more jobs could be usefully created in constructing the infrastructure and buildings that are required to make a low carbon, safe and secure, peaceful and prosperous future possible.

Envisioning a Better World: Tesla, Cobalt & Kolwezi

On Wednesday 24th November 2021 the BBC Panorama documentary was entitled 'The Electric Car Revolution: Winners and Losers'.[58] In it Darragh MacIntyre investigated Tesla's cobalt supply chain. The Kolwezi region of Congo is by far the biggest source of cobalt and Darragh MacIntyre went there to talk to the miners, their representatives and the Sisters of the Good Shepherd who are working to offer help to the local community. Tesla is now a multi-trillion dollar company and Elon Musk the world's richest man. The corporate giant Glencore pays so-called royalties to a ghastly exploitative middle man and they all profit handsomely. Meanwhile the population of Kolwezi lacks even basic necessities like food, clean water or sanitation. Big corporate controlled mines operate cheek by jowl with chaotic artisanal mines. One co-operative of 15,000 artisanal miners was shown in the programme. They work in appalling conditions for a pittance, and death and injury are common as tunnels frequently collapse.

This programme caused me to ponder, how could things be different? In this book I have frequently used the phrase that 'everything needs to change'. So if the kinds of changes envisaged in my Global Green New

Deal were enacted how would this impact on Tesla, Glencore and the cobalt miners of Kolwezi?

Elon Musk would no longer be a billionaire and his Mars mission would have been abandoned. Tesla might still be making cars, but they would be lighter weight, smaller and slower and therefore very much more energy efficient and mainly for shared use rather than for individuals to own. Workplace democracy might mean that Musk played a leading role in the company, but he might well just be an equal member of a co-operatively run company. Tesla would be working more for the public good than for shareholder rewards. Similar changes would have democratized Glencore. The corrupt middle man featured in the programme might well be in jail.

The situation on the ground in Kolwezi would be profoundly different. The wealth taxes that would have limited the ambition of Elon Musk would have helped fund huge changes benefiting the poor of places like Kolwezi. Taxes on extractive industries like cobalt mining might have acted to decrease demand and increase the proportion of cobalt recycling, but we would still need considerable quantities of cobalt as we transition away from fossil fuels and into renewables, batteries and the like.

The Glencore mine would have been merged with the artisanal miners' co-operative and be linked up to, for example, a mining operation in Sweden that might have the best operating conditions, with fewest accidents, most careful environmental procedures and highly efficient operations. The workers would all be reasonably well paid, but in a more modern system there would be far fewer of them. There would of course be many new opportunities. Local government in Kolwezi would have been vastly improved, major investments made in water and sanitation, health and education. Agrivoltaics would be at the heart of a revolution in food and electricity provision in the Kolwezi region. A universal basic income would provide a social underpinning, and of course there would be many new jobs and apprenticeships, education and training. Mining would provide work for some, but there would be a great deal more choice of jobs and roles and a generally vastly improved standard of living. The Sisters of the Good Shepherd might still be active in the area, but the local people would be less reliant on their help and very much more empowered to help themselves.

Congo and the USA would have diminished as political entities, as political and economic power was decentralized into our network of 39,000 or so local communities, co-operating together and assisted by a greatly empowered global networking structure. The Reno and Sparks area of northwest Nevada, where the Tesla gigafactory is located, has a population of about a quarter of a million, and Kolwezi about half a million. The lives of people in Nevada and Kolwezi are now very different, and their knowledge of each other no doubt very poor. What if they were linked together, where people would travel to do a year or two working each other's communities, travelling of course by hydrogen fuel cell powered ships rather than aircraft? Might the miners of Kolwezi have some kind of seat on the board of Tesla and the folks from Nevada a corresponding interest in the mines of Kolwezi? Might both regions' school systems be usefully sharing experiences via Zoom and the children of both communities learning about what they have in common and what opportunities they might have to develop new friendships, new projects and a greater sense of global solidarity?

Chapter Four

Food, Farming and Biodiversity

Introduction

The fundamental question here is, 'Can we feed nine or ten billion people a much better and healthier diet while also restoring global biodiversity?' My answer is an emphatic 'Yes!' with of course the usual caveat, that 'Everything Needs to Change'. There are many changes to diets, farming systems and social and economic justice which if applied as envisaged in this book could see more food, and better quality food, produced off a very much smaller area than is now used by global agriculture. These myriad changes are vital in order to overcome poverty, hunger and starvation. They would also overcome much of the obesity and general lack of mental and physical wellbeing so prevalent in the world today. These same changes are also vital to allow the natural world the space to recover. Rewilding could take place over vast tracts of land and water with enormous benefit to humanity and to the other species with which we share this unique planet. Much rewilding could take place within our productively farmed landscape and in our urban areas. This would also suck a lot of carbon dioxide out of the atmosphere and down into the soil, which, combined with the broad range of changes advocated in previous chapters, represents the best hope humanity has to avert climatic catastrophe.

The changes required are many and complex. Some have been covered in the previous chapters. The envisaged Global Green New Deal would involve changes to political organisation, wealth distribution and taxation, the use of energy, materials and technologies. All these changes would have knock-on effects in terms of food, farming and biodiversity. We have seen how many trillions of pounds, dollars or Euros could be raised. We have briefly discussed the formation of The Global Trust for

People and Planet as a way of channelling these vast sums of money into projects that would be socially and ecologically beneficial. As many of these necessary developments are not currently profitable, in existing systems they are not implemented, or not at anything like the required scale. Much of this revenue could be invested, via The Global Trust for People and Planet, into a huge range of model farms to demonstrate the techniques that I am advocating. The possibilities for improvement to the global systems of protecting nature, and for providing humanity with a good diet, are almost limitless, as we shall see later in this chapter.

First we will look briefly at our current systems and what is wrong with them. The main part of this chapter is about better ways of managing our land and seas, and of changes to food production and distribution. The situation is of course complex and no single mantra like 'vegan is good: meat is bad' is actually very useful even if a reduction in global meat consumption and an increase in fresh fruit, vegetables, nuts and seeds would be very beneficial. Of critical importance is always the detail of how any food is produced, how this affects the ecology of the soils or seas, and who reaps the benefits of its production, either nutritionally or financially. We will also look at how my concept of a Global Green New Deal might be applied so as to ensure that the desired goals of ensuring a better diet for all, restoring biodiversity and sequestering carbon can all be achieved.

What is Wrong with the Current System?

So much is wrong with the current systems of food production, distribution and consumption it is hard to know where to begin. We produce enough in sheer volume to feed everybody, but global inequality means still many people go to bed hungry, while huge quantities of food are wasted. Obesity has rapidly become a global problem. It is often found in people who simultaneously have vitamin and mineral deficiencies, showing how over consumption of the wrong foods co-exists with under consumption of healthy and nutritious foods.

Over processed foods, laden with sugar, fat, and a huge range of chemical food additives that provide little or no nutritional value are currently cheap and easily available, whereas fresh, local, organic produce is often more expensive and less readily available. My proposals for a Global Green New Deal would seek to reverse this by taxing or banning many of these unhealthy foods (and the technologies and corporations that underpin them) and ensuring that the best systems for producing good healthy food are supported by generous new investments. The field has to be tilted so that doing the right thing becomes the cheapest and easiest option. That applies to every aspect of our food production, distribution and consumption. The scale of change is immense.

Industrial systems of meat production such as the American beef feedlots, the poultry broiler houses so common here in Herefordshire, and intensive pig units, are wrong in so many ways. The carbon emissions of such systems tend to be very high, the standards of animal welfare very low. Hormone implants and antibiotic misuse have many associated problems, and the animals' manure is often so concentrated and contaminated it pollutes rivers and streams instead of feeding the soil as it would in a well managed and diverse pasture based system. Some people have suggested taxing meat, and this may be useful, but seems a very blunt instrument, when it is these systems of industrial meat production that are so damaging, rather than meat production as such. Standards of animal welfare need to be improved.

Growing annual grain crops, transporting them huge distances and then feeding these grains to animals is a very inefficient use of resources: much better to let people directly eat the grain, and livestock to graze pastures. Not all pasture based systems are sustainable, but some are, as we shall see later in this chapter. Overall a reduction of perhaps 70 or 80% in global meat consumption would be a very good thing. And all the remaining meat production should be of the very highest standards of multi-species pasture-fed regenerative and agroecological farming. It could be argued that this meat production is almost a by-product of soil based carbon sequestration and kick-starting the microbial life in soils, so vital for restoring a healthy environment. Later in this chapter, and in chapter 5, we shall look at these best systems, and how they could be built upon and further improved to be a very beneficial part of restoring biodiversity and feeding everybody on Earth a healthier diet.

Globally our seas and oceans are in a parlous state. They are full of plastics and a vast and complex range of other pollutants. Overfishing has resulted in declining fish stocks. Trawling and dredging has decimated what should be pristine sea-bed ecosystems. Oceans are warming and acidifying at an alarming rate due to continued carbon emissions and this is changing ocean currents and further threatening fish stocks. Ice caps melting, combined with the expansion of seawater caused by Global Heating, is leading to sea level rise threatening coastal and low lying communities. Warming oceans generate more frequent and more damaging hurricanes. The loss of mangrove swamps and coral reefs further exposes people to the dangers of hurricanes while also losing vital ecosystems which are the breeding habitats of many species of fish. All this is well known.

Reversing it all is possible but it requires vast and concerted global action. Laws, and enforcement of such laws, need to be much stricter. Pollution of all kinds can and must be reduced and then eliminated. Fishing techniques such as dredging and trawling which devastate seabed ecosystems should be banned. Marine reserves need to be created over perhaps 30% to 50% of the world's oceans. None of this means that no fish should ever be caught, or that no industrial development is made at sea, but it does imply very much tighter regulation and control, with the preservation of valuable ecosystems taking precedence over the reckless pursuit of short-term profit.

Just as our systems of meat and fish production need to change, so too do the systems of producing grains, fruit and vegetables, nuts and seeds. Huge fields of wheat being cut by great rows of combine harvesters somehow typify the mid-twentieth century view of progress. Increase in grain production more or less matched population growth over many decades, but at terrible cost. The more that fields were ploughed, and had ever more artificial fertilizers, herbicides, insecticides and fungicides applied, the greater the damage to the soil. Most arable and horticultural production in much of the world became very dependent on massive doses of inputs, resulting in chemical residues in foodstuff, pollution of streams and rivers, and, as mentioned, damage to the soil. These grain monocultures became ecological deserts, devoid of microbial life in the soil, devoid of insects, birds and mammals.

Nearly all commercial development of genetically modified crops has been by multinational corporations which are compounding the problems of modern agriculture. The seeds are designed to be used in conjunction with broad spectrum herbicides such as Roundup. By patenting the DNA of seeds they are appropriating thousands of years of seed development by countless people and privatizing it. It concentrates political and economic power in ever fewer people's hands. All these trends are of course bad in themselves, but what is also of note is that increases in production created by the use of GMOs tend to be very small, short term and built on a model that is wrong in so many ways. There are many other ancient and modern technologies that have vastly more potential to increase global food supply but which get very much less media coverage or political support. Greenhouses, solar powered desalination, precision farming, agroecological methods and any number of other technologies and systems have far greater potential to feed humanity, and to do so sustainably, than GMOs which in general just amplify many of the problems of our current system.

Our systems of farming are the major cause of the loss of biodiversity. Professor Andrew Balmford argues for a concept called 'landsparing', using less land to grow more food so more space can be left for nature to flourish and carbon to be sequestered.[59] The vision of farming that I am advocating throughout this book is a very particular form of landsparing. Balmford's vision is located very much within existing systems of capitalism, and he happily advocates technologies I would argue are far from ideal. My arguments see the necessity of system change to create global economic justice, without which species preservation seems impossible, even of our own species. While he would assume field scale monocropping normal and good as long as it is productive, I would say that there are always better systems.

My advocacy is for complex organic polycultures, growing multiple crops and livestock on the same land. There is a whole emerging plethora of words describing the systems I am advocating, including agroecology, agroforestry, regenerative farming, permaculture, community supported agriculture, organic market gardening, protected cropping, greenhouses, agrivoltaics and many others. Each term implies a different meaning and all of them are useful concepts and there are great examples of each, but none quite captures all of what I am advocating. As my advocacy is couched within the concept of a Global Green New Deal, it assumes the availability of massive financial investment and human input that none of these systems can imagine. Most of these systems are being worked out daily in real world farming practices, and are therefore constrained by the economic realities of the here and now. I look at what they are doing and I am impressed, but always thinking how could this be better if they had more money and human labour to put into this farm or this project. Most practising farmers and smallholders are not thinking about how their farms could be improved if they had a few million, or a few billion, pounds to invest. That is a luxury I do have writing this book. Asking that question opens up extraordinary possibilities. It is those possibilities that I want to explore.

Self-sufficiency, Scale and Technology

The systems of farming that we now call conventional farming are far from conventional if viewed from a longer term or more ecologically informed perspective. Excessive applications of chemicals and ploughing have damaged the land and trawling and dredging have damaged the seabed. Let us refer to this globally productive yet destructive system as the dominant system, because it has indeed come to dominate most of the planet. Also this dominant system seeks to dominate nature, when working much more closely with nature is at the heart of all that I am advocating.

Opposition to the dominant model has been around a long time, and important on every continent and country. Small farmers and peasants, indigenous cultures and many others have long resisted each stage of the development of our modern destructive, yet productive, system.

The Soil Association was founded in 1946 to develop a very different model, founded upon sound ecological principles.

Back in the 1970s John Seymour captured the mood of many early environmentalists keen to develop these organic ideas and practices. Many of us gave up careers in the urban areas of the UK and moved to the countryside with the intention of growing as much of our own food as possible, and in many cases also renovating old ruined cottages, and sometimes living off-grid. Some of us saw ourselves as pioneering a post industrial and rather utopian lifestyle, rejecting consumerism, careers and much else. Many of us became active in the peace and environmental movements and the early days of the Green, or as it was then, Ecology Party. This move to the countryside was dependent on a supply of cheap property with land. The 1980s property boom put an end to it.

This movement was part of a long tradition that portrayed rural self-sufficiency as some kind of ideal, and that growing fruit and vegetables in gardens and allotments would be how a lot more food could be grown. In times of emergency such as the Second World War the dig for victory movement succeeded in raising the proportion of home grown food both quickly and quite dramatically. Some people believe a return to this garden and allotment scale of growing is inevitable within the context of the climate emergency. I am not so sure. It is hard work, time consuming and seen in financial terms it saves very little money relative to the time it requires, but of course it has many other benefits in terms of physical and mental health, and metrics such as how economically worthwhile it is can change rapidly as the economy changes. With an urbanising world, and most people busy with other activities and with the many advantages to doing things on a larger scale, I think it is unlikely to ever again provide a very large proportion of humanity's food, but it could do.

It is to farms that we must look to feed most of humanity. Household level production will always be limited in terms of the range of crops grown, species of livestock kept, and scale of investment made. The complex polycultures I am advocating can be pioneered on family farms, but they will usually need a few extra pairs of hands, and ideally a good team of people. If farms are additionally to be centres of scientific research, education and training, of ecosystem restoration and carbon sequestration, those teams of people might need to be quite big, perhaps ranging from a couple of dozen staff up to a couple of hundred, or even a couple of thousand.

Self-sufficiency remains a worthwhile goal, but instead of thinking in terms of the household level, or the national level, I think there are advantages if we think of applying it at the level of the community. Most of our notional network of communities averaging about 200,000 people might seek to grow as much of their own food as seems sensible within the very variable contexts that these communities exist. Most rural communities should be able to produce a very high proportion of their own food, plus a good surplus. Obviously the more intensively built upon areas will be able to produce a smaller proportion, but with multi-storey rooftop greenhouses above supermarkets, offices and factories they might be able to grow a sizeable contribution to their diets.

Advocates of self-sufficiency have also tended to reject much technology, typified by John Seymour whose ideas were rooted in new applications of essentially eighteenth and nineteenth century technologies. For back garden, family scale vegetable growing very little technology is needed, but to feed large urban populations many new technologies are of use. The Dutch have raised their productivity with greenhouses, drones and precision farming, so applications of water and chemical inputs can be much more accurately and sparingly made. They are one of the leading countries in seed development and production, and do so without using genetic modification. They are a small densely populated country yet are the world's second largest food exporting county, and we have much to learn from them as we shall see later in this chapter.

*

Applying the Global Green New Deal

Agriculture's dependency on the damaging use of chemicals and machinery came about as these inputs became cheaper than human labour over the last sixty or seventy years. It is often argued that by stopping using so many inputs and converting to organic systems yield per acre decreases and therefore it cannot hope to feed a growing population, let alone leave space for rewilding. This line of argument assumes a like for like swop, so one acre of organic wheat would replace one acre of non-organic wheat. What I am proposing is a radically different vision.

At the heart of my vision is a shift away from any kind of monoculture towards complex polycultures of many species of plant and animal on the same land. To farm like this requires many more people per unit of land, but it can produce more food and more biodiversity on the same area of land. We have previously looked at some Pigovian taxes, taxes on things which do damage to people and to the planet. All the chemical inputs including herbicides, fungicides, pesticides, artificial fertilizers and agricultural diesel should be heavily taxed or their use restricted by ever tighter regulation and outright bans.

We need more highly skilled workers on the land, well versed in the emerging systems of production that are more ecologically regenerative and therefore also more sustainable. The Global Trust for People and Planet could buy land to use as model farms and as training centres. We looked briefly at how this Global Trust for People and Planet might be organized. At the heart of this was the recruiting of millions of people to do the work needed to ensure humanity has a future, and to further ensure that it is a future that is better in a vast array of different areas, where biodiversity is once again increasing, climate change has been reversed, and hunger, poverty and many other problems eliminated. The changes I am proposing to global agricultural systems will help heal all these problems.

Currently there are millions of people fleeing war, drought and poverty. There are also millions of people unemployed or in soul destroying, poorly paid and insecure employment. Automation will only add to the masses of unemployed, and the necessary contraction of industries that are part of the problem will cause many more to lose their jobs. The number of interesting, worthwhile and enjoyable jobs that could be created by the changes that I am advocating for the food and farming sector might be in excess of one hundred million. It might be that a billion new jobs are created. The possibilities are indeed endless.

We have discussed changes to taxation and subsidies aimed at tilting the field toward making these better systems of food production cheaper and easier for people to access, and so more widely available, and we have discussed getting many more people actively involved, working to develop these better systems. At the heart of all these changes I am proposing a global network of model farms.

A global network of model farms

In the section on governance we looked at a decentralized model and imagined a network of about 39,000 local communities, each of about 200,000 people, totalling the 7.8 billion of us alive today. Perhaps all, or very nearly all, of these communities would have a model farm adapted to their local conditions. Many communities might want a whole network of model farms across their region.

These model farms would seek to demonstrate sustainable systems, and to continually improve them. They would play a major role in education and training, and in food production. An annual investment of about, on average, 25 to 30 million pounds for each of these 39,000 communities would work out at about 1 trillion pounds per year. In Chapter One we envisaged a global annual budget of 5 trillion dollars being invested in this sector. In today's exchange rate 5 trillion dollars would be 3.78 trillion pounds. So let us imagine what our Global Trust would do with something between 1 and 3.78 trillion pounds per year to be invested in the food and farming sector.

With this they could invest in land, buildings and recruit staff. Every year they would take on a new cohort of apprentices and students learning the many skills required for these kinds of innovative farming projects. As production on these farms increases the level of subsidy could probably be reduced, but not eliminated. Their role in education and training, in improving diet and health, in sequestering carbon and enhancing biodiversity would justify a considerable level of ongoing financial support. Also each year they would be expanding into new geographical areas, and initiating ever more ambitious projects.

Currently most farms employ few people and the farmers are often old and struggling to keep going, doing more or less what they have been doing for many years. They lack the time and energy to put into reading, research and visiting other more innovative farms. They also often lack the capital to invest in new systems. Many farmers are burdened with debt. As farming has employed ever fewer people, isolation, depression and suicide have become all too common in many farming communities around the world.

Those farms that employ a lot of people tend to recruit casual workers, in the UK usually young east Europeans, in the USA often Mexicans. They don't invest in the education and training of this workforce. What I am proposing is a system where young people, and especially poor young people, wanting a career in horticulture and farming are seen as perhaps the most important resource on the planet. They should be invested in, educated and trained, and given long term secure access to land or to employment. The network of model farms that I am proposing would have teams of people working together where all are valued, and properly remunerated, members of a team. We may come to see that teams of farm workers and teams of research scientists have more in common than we could possibly imagine. Most of our network of farms would have people doing post graduate research on topics such as the role of mycorrhizal fungi in soil and plant health, alongside people learning skills such as welding and of course some people might be doing both.

The vast majority of our global network of model farms would also be producing some energy on-site. Nearly all would have solar panels for producing electricity and hot water on barn roofs, and many would have field-scale solar, and all this solar technology would be integrated into crop production and biodiversity enhancing land use in various agrivoltaic systems. Some of our model farms would have wind turbines, anaerobic digesters and some would have heat pumps and geothermal systems. Farm machinery would largely be battery electric or hydrogen fuel cell, probably with on-site electrolysis for turning surplus electricity into hydrogen. I will explore all this in more detail later, but while I am discussing the training and education of workers on these model farms it is important to note that many farms would have teams of people able to design and construct systems of energy generation and storage.

At the end of the second chapter I talked about the Global Trust for People and Planet taking an annual intake of say 70 million sixteen year olds and starting them off on a long period of education, training and apprenticeships geared toward helping heal the many problems afflicting humanity. If this super apprenticeship was ten years long I noted that this would generate 700 million 16 to 26 year olds, the older ones being very highly trained. The Trust would also offer training, education and work to older people. As I have noted the Trust would not be a vast monolithic organization but rather a lose association of numerous projects each with a high degree of local autonomy working toward broadly agreed global goals. This workforce would be involved in building infrastructure such as water and sanitation systems, tram systems and cycle paths, schools and hospitals and in staffing them. Perhaps the network of model farms that I am proposing would employ and offer training to more people than any other sector. What might a better system of global agriculture look like if we invested more money in it and had several hundred million highly motivated and trained people working in it? What would they all do? Perhaps the task of rebuilding soil fertility, globally, acre by acre across most of the farmed landscapes is the most important and first step.

Soil and Compost
On a global scale, it has been estimated that soils hold more than twice as much carbon (an estimated 1.74 trillion US tons) as does terrestrial vegetation (672 billion US tons). Raising the organic carbon content

of the soil through applications of manure and compost and good soil management sequesters carbon from the atmosphere, which will be of critical importance as we reduce and then reverse the damage done by global warming. Increasing organic matter, most of which is carbon, increases the ability of the soil to allow water to percolate into the ground, so helping to prevent both flooding and drought. Healthy soils, with plenty of organic matter, exhibit an extraordinarily complex range of ecosystems, with many species of microbes, fungi, insects and worms each playing a role in ensuring long term soil health and productivity, and forming the base of food chains for birds, mammals and the rest of a well functioning ecosystem. The collapse in global biodiversity has to a large extent been caused by the systems of farming that have dominated agriculture for most of the last sixty to seventy years. We need to farm differently, and a wealth of alternatives exist which show that we could feed everybody an excellent diet while sequestering carbon and restoring biodiversity.

Over the years I've been inspired by visiting and reading about many pioneering projects and farms. I've worked briefly on a few of them and have been a keen fruit and vegetable gardener for many years. What I have learnt forms the basis for what I am proposing here. I have seen plenty of examples that prove that we can grow more food, and better food, than is currently the case. We can feed all of humanity a better diet, and we can do it while restoring the biodiversity of the natural world.

Our network of 39,000 communities, each with their model farms, would all be co-operating and sharing knowledge about restoring and enhancing soils. Stopping the damage done by excessive ploughing and chemical inputs is the first step, then blocking erosion gullies and planting hedgerows across the line of slope. There are so many more steps. Establishing complex polycultures with trees, perennial and annual crops with many species of livestock integrated into the mix and plenty of compost added to this all will help build soil carbon and soil fertility. Other additions might include the digestate from anaerobic digesters, green manures and plenty of legumes in the rotations to increase nitrogen levels. Biochar has enormous potential to improve soil carbon storage and soil fertility, but it needs lots of research and experimentation on many different soil types and in many different climatic zones: in some it will be hugely beneficial, in others very much less so.

Adding rock dust has many potential benefits to the soil, and like Biochar needs lots of research and experimentation in different soils and climates. Basalt as it weathers absorbs carbon from the atmosphere, adds alkalinity and remineralises the soil.[60] It looks promising as a way both to enhance soil fertility and to sequester carbon from the atmosphere, and as the water flowing from these fields enters the oceans it also helps neutralize ocean acidification. Our network of model farms would facilitate this research, development and deployment of all these techniques of soil restoration and enhancement and on a truly epic scale.

I have argued that a shift in diets toward more super fresh, local, organic fruit, vegetables, nuts and seeds would be good for both people and planet. Also I have argued that a reduction in the number of farmed animals, and a corresponding reduction of meat consumption of perhaps 80% would be very beneficial. So, perhaps paradoxically, I want to start by singing the praises of some meat production. I have advocated the closure of all factory farms, such as the American beef feedlots, Herefordshire style poultry broiler houses, and intensive pig units. I would also like to see a massive reduction of monocultural sheep grazing so common today in the British uplands. I have argued that farm animals should be integrated into complex polycultural farming practices. Instead of farms specializing in producing just one or two crops or species of livestock I envisage nearly all our global network of model farms producing a vast range of all manner of produce, probably several hundred different products coming off each farm, in some cases probably several thousand.

Most farms would have some livestock of many species, but not too many of each, and they would be well integrated with both annual and perennial crops. In the eighteenth and nineteenth centuries most farming was indeed mixed along roughly these lines. What is needed is to extrapolate from this old model of mixed farming and build in many new ideas and techniques and vastly greater diversity of all manner of species of plant and animal.

Meat

In the popular imagination meat is now seen as bad from a climate change point of view, and indeed this is the case with all industrial meat production. Ranching has been responsible for much of the ecological onslaught on the Amazon basin, and poorly managed livestock systems are responsible for much damage to many habitats and ecosystems. Livestock farming does not have to be like this. It can be done well.

Under the best pasture based systems, of course, cows still burp and produce methane, a powerful greenhouse gas, but this is more than offset by the carbon sequestered in the soil. Let us look at one of the best farms of this type.

White Oak Pastures in Bluffton, Georgia, USA, is a fine example of regenerative pasture fed livestock systems. In a study by Quantis, White Oak Pastures beef was shown actually to be carbon negative, achieving net carbon sequestration of 3.5 kilos of carbon dioxide per kilo of meat, making it a lower carbon food than soybeans or vegan burgers.[61] They are a very unusual farm, in some ways unique. They rotate cattle, pigs, sheep, goats, rabbits, hens, turkeys and ducks and it is this huge diversity of livestock that helps make the system work. Each species eats differently and contributes a different type of manure, so maximizing the diversity of dung beetles and building soil based biodiversity. The livestock are grazing on species rich permanent pasture. Rotating and mixing livestock breaks the pathogen cycle, so poultry eat any parasites within the cow's dung, and the land has plenty of time to recover and the sward to grow back before the livestock return.

They are the only farm I know of that has on-site abattoirs for both red meat and poultry, and the only farm I know of that really has a zero waste system of utilizing the entire animal. The entrails are used for raising maggots to feed to poultry, and then made into compost. The hide is tanned and made into leather. Every last part of the animal is used. They market all their produce directly to consumers, and over recent years have expanded their acreage from 1,000 to 5,000, buying and renting more land to expand their operation and keep up with demand. Each year they are planting more trees into their pastures to provide shade and forage for the livestock, to improve the ecology and to provide additional income streams. Pecan nuts do well in their climate and are a growing aspect of the farm. In 2012 they started a small, 3 acre, experimental vegetable plot growing a huge diversity of fruit, vegetables and herbs using an organic zero tillage system.

White Oak Pastures is predominantly a meat producing farm, and I have argued that a lower level of meat production would be a very good thing. Any form of meat production requires a lot of land. Vastly more nutritional gain can be had per acre by growing fruit, vegetables nuts and seeds, but if we are to raise livestock White Oak Pastures does provide a very good model, and the gains in soil carbon sequestration and fertility are to be applauded. Many farms are now developing similar, if somewhat less ambitious, systems of regenerative pasture fed livestock farming. There is much variety within these systems, with some such as White Oak Pastures being mainly a meat producing farm. At the other end of the spectrum we might place the Knepp estate in Sussex, for example, which does raise livestock to butcher, but really more as a by-product of the UK's most successful rewilding project. Other farms, like that of George Young of Fobbing in Essex, that have been focused more on grain production, are adding lines of trees and a small herd of red pole cattle to create a more balanced agroecological system.

I would like to see very much more research and development on how best to manage livestock to improve degraded soils and create a kind of ecological succession. Most land that has been under arable farming for decades now has very much less organic matter in the soil than it ought to have. Putting such land down to White Oak Pastures type livestock farming would rejuvenate the soil, building soil fertility and storing carbon. After a few years the next stage in the ecological succession would be to gradually introduce more fruit, vegetables, nuts, seeds and grains and reduce grazing. White Oak Pastures have their three acre zero tillage vegetable garden on their now super fertile soils. I would love to see this expanded to thirty acres, and if this was successful, then to three hundred acres. The livestock would move on to other farms which had damaged soils.

My envisaged Global Green New Deal sees meat consumption ideally reducing by about 80% and of all of the remaining 20% to be in these types of farming system as pioneered by White Oak Pastures, the Knepp Estate and George Young of Fobbing. The model farms network would seek to integrate many other aspects, so not only meat would be produced, but also, many other crops, notably fruit, vegetables, nuts and seeds, and also plenty of nutrient dense grains. If we are to eat less meat, fish and dairy produce, it will be increasingly important to make sure we eat plenty of other nutrient dense foods. There is tremendous scope to increase the role of nuts, seeds, legumes, and nutrient dense grains into the human diet.

Grain

The great grain monocultures have helped feed a growing population, but with every year that passes the soils on which these systems are practised deteriorate. Excessive ploughing and harrowing destroys soil organic matter, that vital carbon store and aid to fertility and to soil structure. It exposes the land to erosion and makes droughts and flooding events more likely. At present huge quantities of these annual grain crops are grown to be fed to livestock, or for putting into anaerobic digesters to produce gas and subsequently often electricity, or are grown to be turned into bio-ethanol or biodiesel.

It seems to me that the only ethical use of growing grains is for direct human consumption. There are better ways to produce energy for electricity, heating, cooling, transport and industry, as we saw in the last chapter. The regenerative and agroecological farming systems that I am advocating need very little or no grain to produce a great variety of livestock. The total area given over to grain crops globally could be considerably decreased if it was all, or very nearly all, for direct human consumption.

Instead of having great monocultures of grains we should integrate our grain cropping into agroecological systems. So, for example, strips of grains can be grown between rows of fruit and nut trees in an alley cropping system, and be rotated with a vast range of legumes, vegetables, fruits and livestock.

Wakelyns is a 56 acres farm in Suffolk which has pioneered an organic system of alley cropping since the 1990s. Rows of a huge diversity of trees, mainly fruit and nut trees, are grown with strips of pasture or cultivated land between. These alleys are 12, 14 and 18 metres wide, and crops and livestock rotated along each in turn. The range of crops is huge including lentils, oats, peas, chia, camelina and many heritage varieties of nutrient dense wheat. The rows of trees serve many useful functions apart from the great diversity of crops, being reservoirs of mycorrhizal fungi. They provide shelter to the arable crops and livestock and also allow water to percolate into the soil so reducing water run-off and soil erosion. This year, 2021, they are restarting growing vegetables through an organic box scheme.

George Young farms 486 hectares, or 1,200 acres, in Fobbing, in south Essex. He is a pioneer of zero tillage systems for growing an extraordinary range of crops, including even more varieties of nutrient dense heritage wheat than Wakelyns, as well as buckwheat, lentils, linseed, hemp, sunflowers, millet, tiger nuts and heritage corn. He is planting a vast array of types of fruit, nut and timber trees in rows with cropping alleys between, much as they do at Wakelyns. At any one time he has about 20% of his land down to herbal leys where he grazes red pole cattle, which he integrates into the cropping rotation with all the grains and legumes. He mills some of his own grain on site and works with a small local firm to make pasta from some of these nutrient dense grain crops. Much of the rest he markets directly himself or via Hodmedod's.

George Young employs one full-time helper and his dad helps a bit. He also has a few occasional volunteers helping, but as George himself is also working on the processing, marketing and distribution of produce, his farm has probably less than two full-time people actually working on the farm looking after all the livestock, crops, tree planting, fencing and the like. I wonder what he would do if offered a few dozen apprentices from our envisaged network of model farms? What if we used his farm and lessons from Wakelyns, White Oak Pastures and other farms and constructed a model that might be expanded and applied in the very varying contexts of our global network of model farms? We shall explore this, but first let us consider other elements to add to our envisaged network of model farms.

Fruit, vegetables, nuts and seeds

Having a diet rich in fruit, vegetables, nuts and seeds is what most health advice recommends. Colin Tudge in 'Feeding People is Easy' shows this to be the basis of a healthy diet and many of the world's great gastronomic traditions, and also the basis for producing more nutritional value per acre than under any other farming system. Utilizing greenhouses and irrigation can considerably increase this production per acre, and with some of the best types of greenhouses this highly productive horticulture can be practised in almost any climate on Earth.

It is amazing how much food can be grown on a small area. In my urban garden every day of the year my wife and I have a range of salad crops and several types of vegetables ready to harvest. For about six months of the year there is an abundance of fresh fruit, with plenty to freeze to see us through the winter and spring. In my average sized town garden I grow rhubarb, strawberries, raspberries, loganberries, blackberries, blueberries, jostaberries, gooseberries, blackcurrants, whitecurrants, redcurrants, grapes, cape gooseberries (physalis) and then there are the trees: apples, pears, cherries, plums, greengages, damsons, peaches and apricots. We also grow elder trees for elderflower and elderberry cordials. We used to grow hazelnuts but decided we didn't have the space. There are many more species I'd love to grow but we don't have the space, the time or really the need for any more fruit.

The UK imports 84% of all the fresh fruit that we eat.[62] This means that only about 16% is grown in the UK. This is such a bizarre state of affairs. We could reverse these ratios with many positive outcomes. Just to take one example. Plums that are grown in distant countries and shipped to the UK have to be picked under-ripe and therefore of poor flavour and poor nutritional content. Really ripe plums are delicious and nutritious, but need to be eaten as soon as they are picked. It is almost impossible to transport, store and retail them in the kind of model that supermarkets and greengrocers use, so most people in the UK have never eaten a properly ripe plum. Before the days of cheap imports we used to grow more plums than we do now. I recall old and often neglected plum orchards in Herefordshire which now have all gone. In their heyday the fruit would have been harvested and sold locally, or bottled or made into jam for longer term storage. Plums, like so much fruit and vegetables, should be grown close to where people live, and picked and eaten on the same day, or failing that the very next day.

My proposition for a global network of model farms located in pretty well every community on Earth would seek to grow as wide a range of fruit and vegetables as possible, and for these model farms to be as far as possible within a fifteen minute walk or cycle ride of where everyone was living. This might mean that some of the more sparsely populated areas would need dozens of smaller dispersed gardens and greenhouses. Everyone should have access to the very best organic and super fresh fruit and vegetables within a fifteen minute walk or cycle ride from where they live, and sold at prices everyone can afford. Nobody should be without easy access to healthy food: it should be seen as a fundamental human right. The farms located in inner city areas would probably concentrate on growing salads, herbs, and those fruits and vegetables that are best picked and eaten super fresh. Most nuts, seeds, grain, meat and some types of vegetables would still need to be grown where more space is available in the countryside, but in as close proximity to the people who will eat the produce as possible.

The cost in terms of food miles of current Western diets is outrageous. This of course results in carbon emissions, air pollution, traffic accidents, poor food quality, and as larger corporations can better exploit international markets the concentration of land holding into large multinational corporations, often resulting in poor working conditions for those employed in the system. Food miles could be massively reduced, certainly by at least 90%, with huge reductions in all the problems associated with excessive food miles.

Eating a wider variety of fruit and vegetables will be important. Greenhouses and other forms of protected cultivation provide an excellent opportunity to extend the range of crops grown and the length of the growing season in any given location. They should therefore greatly help reduce global food miles, and will be a vital component of our network of model farms.

Greenhouses and Agrivoltaics

Greenhouses, polytunnels and various forms of crop protection have the potential to dramatically increase the amount of food grown per unit area. This fact is demonstrated by the example of Dutch agriculture. Holland is a relatively small, densely populated and industrialized country, yet manages to also be the world's second largest food exporting country, ranked by value of its exports. It has done this largely by pioneering the uses of greenhouses and crop protection. Not all of this is sustainable, for example the use of fossil gas for heating and the often excessive use of chemicals. However there is much to learn from what they have achieved and how they may be improving their ecological impact. At the forefront of these improvements is Wageningen University and Research Centre, which has spun out many innovative start-up businesses utilizing greenhouses. Their mantra is to continually reduce the inputs of fuel, water and chemicals while increasing production.[63] In the Westland area of Holland 80% of cultivated land is covered by greenhouses, and this small region accounts for nearly half of Dutch horticultural production.[64] But it is not just in Holland that innovation in the design and use of greenhouses is taking place.

A new greenhouse at Ingham,[65] just north of Bury St Edmunds in Suffolk, claims a world first, taking waste heat from nearby Fornham water treatment works and utilizing heat pumps to concentrate the heat which is then used to heat the 13 hectare (30 acre) greenhouse. The greenhouse also contains a small-scale combined heat and power plant, which, if fossil gas is to be used, is the optimum place and scale, so the waste carbon dioxide can be absorbed by the growing plants rather than vented to the atmosphere. There is also a grid connection, so presumably when electricity is expensive the CHP plant can sell electricity into the grid, and then buy back electricity to run the heat pump when electricity is cheap.

One trend I very much like is the increasing use of greenhouses on the roofs of other buildings. In the USA, Gotham Greens have a number of greenhouses growing salad crops above large supermarkets, with as much of the produce as possible sold directly below where it is grown, so reducing food miles to food yards. They claim to be using 95% less water than conventional farms and 97% less land.[66] Increasingly, growing space will be integrated into architectural design. One example is the Lumen building at Wageningen where university teaching rooms open on to the courtyard greenhouse. Integrating greenhouses into other buildings adds to their thermal mass, meaning that heat is stored within the fabric of the building, levelling out the difference between daytime and night-time temperatures. This reduces the need for other forms of heating.

One project that sadly never got developed was a five storey greenhouse that an organization called Growing Power planned to build in Milwaukee. Its southern side was to be angled to maximise solar gain and light penetration deep into the building, with aquaponics on part of the ground floor and hydroponic crop cultivation above. On the northern side of the building there would have been offices and classrooms, lift shafts and stairs. It seemed like an excellent design concept, and along with some of the Dutch designs could be further developed in the more urban examples of our proposed network of model farms.

Clear photovoltaic panels are now being developed which have the potential to be utterly transformative to both the global supply of electricity and food. As I write, in 2021, initial trials are under way in Holland with Wageningen University partnering with the German solar company BayWa and its Dutch subsidiary GroenLeven.[67] Monocrystalline solar panels are being designed with varying degrees of transparency and are being tested on five different crops at five Dutch fruit farms. The early indications are that excellent crops can be grown, probably better than under conventional plastic polytunnels, as well as generating electricity on the same land. They look like they will be much more durable and lower maintenance than polytunnels. As production is scaled up cost will fall and it is my expectation that they will largely replace most if not all polytunnels. These solar panels may end up replacing traditional glass in greenhouses.

This system of combining electricity production using solar panels with productive arable or horticultural production is known as agrivoltaics. Very small scale experimental systems have been pioneered since the first decade of the new millennium, but it is only in the last couple of years that the scale of investment in this sphere has really started to take off. Germany's Fraunhofer Institute has been at the forefront of research and development in this field, as in so many areas. However, unsurprisingly, it is in East Asia that the technology has been deployed at greatest scale so far. Japan has over a thousand farms and gardens using some form of agrivoltaics. The largest projects in the world are being developed in China's Gobi Desert, and in the chapter on energy I mentioned a1GW solar installation with goji berries growing under the panels. Thousands of hectares of the Gobi are being developed to grow crops and generate electricity on what was previously barren sandy desert.[68] Almost all these agrivoltaic projects use standard solar panels, but as clear panels become more widely deployed they will help improve agrivoltaics, especially so in less intensely sunny locations.

With the proposed network of model farms it would be possible to trial these clear solar panels, which themselves have varying degrees of transparency, in conjunction with every imaginable climatic and soil type, and every conceivable type of agricultural production. No one single technology is going to save humanity, but some will have a big role to play in turning things around, and agrivoltaics is one of those potential game changers. Another of the potential game changers is the seawater greenhouse.

Seawater and Saltwater Greenhouses

One of the most exciting developments in greenhouse design and construction has been pioneered by the British inventor Charlie Paton, with his concept of a seawater greenhouse. This relatively simple technology has great potential to help open up the world's hot dry deserts to very productive systems of food production and to new cities and new forms of economic development. In Chapter Five we will look at these wider ramifications, but first we need to look at the technology itself and its evolution up to the current time.

From 1992 up to the present Charlie Paton and a small team of collaborators have designed and built a number of experimental structures, first in Tenerife, then in Abu Dhabi, Oman, Australia and Somaliland. Each project sought to show how the energy of the sun and wind could be used to pump seawater to a greenhouse where it would be desalinated and used to grow crops. The first three projects, in Tenerife, Abu Dhabi and Oman, used the sun and the wind to pump seawater to the greenhouses, dribble it down an evaporative surface of latticed cardboard, where the water evaporates, and then condenses on cooler seawater filled tubes. The amount of desalinated water was then sufficient to irrigate the greenhouse and leave surplus for other uses. The project just south of Port Augusta in Australia was the first large scale commercial development of the technology. In 2010 they began with a 2,000 square metre pilot, then, under new commercial ownership, expanded the greenhouse one hundred fold, to twenty hectares. Here they added a 36MW concentrating solar power station to drive the system. On these 20 hectares of former desert they now produce 16,000 tonnes of tomatoes per year. This is land that was too dry for any form of agriculture, and now, primarily driven with abundant and free resources like the sunshine and seawater, they have created one of the world's most productive farms.

Recently Charlie Paton and his team have collaborated with Aston University in Birmingham and the Pastoral and Environmental Network in the Horn of Africa (PENHA) to design and build a project in Somaliland. This version is a simpler, lower cost development of the concept, replacing the greenhouse with shade netting. This again is something that has great potential, especially when used in conjunction with a range of other technologies, such as those developed in a project in Qatar.

In 2012 a Norwegian NGO called the Sahara Forest Project, along with other partners, built a one hectare demonstration project in Qatar.[69] They showed the synergistic benefits of utilizing a number of innovative technologies that work well in hot arid places to produce abundant water, food, salt, electricity and algal lipids that can be the basis of many useful products, from food and fuel to fertilizer. The seven key technologies they sought to demonstrate were concentrating solar power, photovoltaic solar power, saltwater greenhouses (based on Charlie Paton's seawater greenhouse concept), evaporative hedges with associated vegetation, halophytes (salt-loving plants) and the production of both salt and algae. The system makes good use of free resources to achieve multiple benefits. So for example freely abundant saltwater is desalinated in the saltwater greenhouse, producing fresh water for drinking, irrigation or other uses. The resulting brine, which could be a pollutant, is instead dribbled over an evaporative hedge, which is essentially a cardboard lattice, producing a cooler, more humid micro-climate between these evaporative hedges where many useful plants can be grown, and the brine then becomes sufficiently concentrated to be put into evaporative ponds to further dry out. The mix of minerals and salts that is left can form the basis of many other processes or products, from calcium carbonate or gypsum which can be used as soil additives, to minerals such as manganese that have commercial value, and of course huge quantities of salt which also has commercial value. Within this combination of technologies other waste streams can be continually recycled. So for example the crop residues from hydroponic vegetable production in the greenhouses can be directly used as a mulch, or composted first, then added to the growing areas between the evaporative hedges in order to build soil carbon and water retentiveness, and allow the increased production of crops in the former desert areas outside the greenhouse.

More recently this Norwegian organization has been working with many partners in Jordan to develop another small project demonstrating some of these same technologies just outside Aqaba.[70]

These seven projects are, to the best of my knowledge, the only examples of the concept of the seawater or saltwater greenhouse (with a range of compatible technologies) that have so far been built. They demonstrate that by utilizing seawater and solar power, abundant water, food, energy and much else can be sustainably created in hot dry desert regions. Each of these projects has been relatively small scale, most below one hectare, apart from the Australian greenhouse covering twenty hectares. Charlie Paton and the Norwegian team have always had the ambition that vastly bigger examples of these concepts and technologies would get built. It is my belief that they are of critical importance in addressing the whole complex array of problems that this book seeks to address. The clear photovoltaic panels currently under development in Holland that I enthused about in the last section could also be used, adding an eighth to this portfolio of technologies. All the ideas and experience being gained from agrivoltaics projects will have synergistic benefits when combined with these seawater based projects. I can think of dozens, perhaps hundreds of other ideas and technologies I would love to add to this mix.

With the kinds of vast investments envisaged in my Global Green New Deal these larger scale and more complex systems could be developed. The Qatar project was one hectare. What would a project with say 10,000 hectares and a few tens of billions to invest look like? How could such a project best benefit impoverished and troubled regions in the Middle East and the Horn of Africa? These technologies could be developed to be the basis for whole new cities and systems of agriculture capable of greatly increasing global food security and having multiple other benefits.

Seawater Farming

Currently most fish farming is done in ways that are pretty awful, such as salmon farming on the Scottish lochs. Salmon, being a fish eating fish, need to be fed on fishmeal, which is ground up fish, and the whole supply chain of this is wasteful. Fish stocks in places such as Peru are denuded in order to feed these Scottish salmon, which because their cramped conditions are riddled with sea lice and in very poor health, yet form the basis for most salmon produced in the UK. By farming plant eating fish such as carp or tilapia the systems could become more sustainable. Better still would be the development of systems designed to harvest a greater diversity of species, with most of them being lower down the food chain.

Bivalves, such as oysters, mussels and clams look to be one of the best sources of animal derived protein.[71] As they are filter feeders they even have a role in bioremediation, cleaning the water in which they live. Their cultivation can produce a lot of food with minimal damage to the environment and with less ethical concerns than the farming of pretty much any other species of animal. Offshore Shellfish Ltd is a company operating in Lyme Bay in Devon, in the process of developing a fully offshore system of mussel cultivation based on ropes suspended between buoys. They plan to harvest 10,000 tonnes of mussels per year, from three sites totalling 15.4 square kilometres.[72] There seems to be good evidence that biodiversity increases within the farmed area. This seems like the kind of aquaculture that should be very much encouraged. Strangely people in the UK do not eat much shellfish. Much of it is exported to the EU. However, given the bureaucratic chaos unleashed by Brexit, exporters have a major hurdle to cross. However it seems to me that this is precisely one of those many kinds of farming that humanity should be expanding, while carefully monitoring the ecological ramifications.

As more offshore windfarms are being built it has been noticed that the turbine towers make excellent artificial reefs which mussels naturally colonize. By stringing chains, or ropes, between the turbines this reef effect could be expanded and mussel farming could be co-developed with North Sea windfarms. Potentially this could become a major source of food, replacing many unsustainable fishing practices.

Many forms of seaweed are edible.[73] The Japanese, Chinese and Koreans traditionally eat the greatest range of them, but we could all eat a lot more of them in the future. Some forms of seaweed grow exceptionally quickly and in terms of producing biomass per acre they are one of the most productive of any plant or animal. This seaweed could become the basis for much food production, and also as a feedstock in composting systems aimed at enhancing degraded soils, or as a feedstock in other agricultural and industrial processes.

Seawater Solutions is a small start-up organisation that is utilizing salt loving plants, fish and shellfish to farm in novel ways that seem potentially both more profitable and productive and also have many environmental benefits.[74] In Scotland they are working with a farmer to transform poor sheep pasture into a salt marsh environment where they are growing samphire, a highly valuable food crop that naturally grows wild on salt marshes. In Vietnam they are working with rice farmers in the Mekong delta, where seawater intrusion due to rising sea levels is threatening rice production. By introducing halophytes into the system they provide the farmers with an alternative income stream. In Bangladesh they are working with shrimp farmers by introducing halophytes with the aim of extracting excess nutrients from the system. It is early days but this looks to me to be another area of farming that could be beneficially expanded, and if incorporated in combination with a number of the technologies being developed by the Sahara Forest Project, Charlie Paton and others, it could have synergistic benefits. In my envisaged network of model farms many coastal communities would have ample funds to push ahead with research, development and deployment of this whole portfolio of technologies and farming systems.

Insects and Fungi
In this chapter we are envisaging changes to humanity's diet. Eating less meat, but better quality meat raised under better systems is one

step. Eating more local, organic seasonal fresh fruit and vegetables, more nuts and seeds is another crucial step. Bivalves and seaweed may usefully become a larger proportion of the global food supply. So too might insects and fungi. Like bivalves and seaweed, insects and fungi have great potential. Some forms are already eaten quite widely in certain countries and hardly at all in others. The humid tropical regions of Africa are the main area where insects have long formed an important part of the diet.[75] In many other countries, such as France and the UK, there is a commonly held perception that somehow eating insects is disgusting. This is a pity. Some people are now seeking to change this and a number of new start-up businesses are trying to farm and market insect-based food products. The potential to raise more protein per acre than with any traditionally grown animal protein is tremendous. Farming insects in modern factory type conditions may well be the way this new industry develops. Currently huge swarms of locusts do terrible damage in places such as East Africa, and the normal practice is to spray insecticides from aircraft in order to kill the creatures. This aerial spraying does much collateral damage. Imagine if we could develop some system of capturing the entire swarm, perhaps using some combination of pheromone attractants and mist netting, we could harvest many tonnes of valuable protein and also save the farmers' crops from destruction. That to me seems another area worthy of research and development investment.

A great diversity of wild mushrooms and fungi have traditionally been harvested in many countries. A growing number of species are now being commercially farmed. Like with seaweed, many species are fast growing, highly nutritious and could form a very much larger proportion of humanity's food supply than is now the case. The fact that some forms of fungi grow best in the dark makes them ideally suited to be grown indoors, perhaps on the ground floor of a multi-storey greenhouse. They could be grown with aquaculture and salad vegetables above, so getting extremely high production per acre, and making most sense in urban contexts where land prices are high and many thousands of people want access to super fresh food.

SYSTEM CHANGE NOW!

Fungi's complex and important roles in the healthy functioning of ecosystems has only recently really begun to be understood. As humanity gets to grips with cleaning up landscapes ravaged by industrial pollution and inappropriate farming systems, fungi may well play a critical role in bioremediation. The productivity of plants and the quality of the produce that they provide is to a large extent governed by the networks of mycorrhizal fungi that are growing in symbiotic relationship with nearly all plants on the planet. Understanding these complex relationships and designing farming systems that best encourage these beneficial relationships seems to me to be one of the key areas justifying large scale funding for research, development and deployment.[76]

Throughout this chapter I have argued the benefits of integrating numerous crops. Alley cropping is one of the best ways to do this. This involves planting lines of trees with a rotation of vegetables, grains and livestock between them. If the cropping alleys are no more than about 18 metres wide, networks of mycorrhizal fungi can extend from the tree roots and connect to the roots of the vegetables, cereals or herbal leys growing in the alleys, to the mutual benefit of all. If the alleys were very much wider than this the mycorrhizal fungi would fail to encompass the whole crop. Developing farming systems that best work with mycorrhizal fungi in differing climates and soil types would be a critical aspect of the work of the model farms network. The beneficial role of fungi in soil creation and restoration is of vital importance. As we have noted, soils are massive carbon stores, and how we manage them will become ever more important as the climate crisis unfolds.

Farm-Free Foods

A start-up company from Finland called Solar Foods has developed a technique of precision fermentation that could, at least in theory, change the entire global food system over the next decade or two. They have isolated a hydrogen-oxidising bacterium that lives naturally in the soil and worked out how to use it to produce the most extraordinarily diverse range of possible foods. Whereas fermented foods such as quorn require plant-based carbohydrates as a feedstock to produce their higher protein food, the Solar Foods product, Solein, does not. It is made essentially from hydrogen (from water) and electricity (from solar panels) and carbon dioxide (from the air). Small quantities of trace elements such as calcium, sodium, potassium and zinc are added.

Solar Foods started production in 2021, and are planning to be producing about 50 million meals per year by 2022.[77] This factory based process is not dependent on soils, or weather or any of the other factors that constrain almost all other systems of agriculture. It is claimed that 20,000 times more food could be produced per acre than is the case with soya. This means that on an area the size of Ohio, or about 1% of the Sahara desert, enough food could be grown to match the entire global food demand. It is unlikely ever to replace all other foods, but even if it were to supply say 10 or 20% of global food supply it could displace much of the most ecologically damaging systems such as American feedlot beef. It could reduce pressure to develop agriculture in areas which would be better left for large scale rewilding. I have used the term landsparing, developing systems that produce more food on less land, so taking pressure off fragile ecosystems and allowing space for nature to recover. Solar Foods seems to be the ultimate example of extreme landsparing. Just to cite two examples, cattle farming in the Amazon and oil palm plantations in Indonesia could cease, and both these magnificent ecosystems could be rewilded and redeveloped in very much better ways.

Initially Solein will be developed as a high protein additive to manufactured foods like pasta. It will probably then be developed into a vast range of products from cultured meats, fish, eggs and milk and replace soya as a feedstock to many other products. Possibly, in theory, it could even be used to produce lab-grown timber.

As ever the real impact of such a technological breakthrough will depend on the right global legislative framework. At the extremes, under a market fundamentalist system it would be patented and all the profits go to a tiny number of people. At the other end of the spectrum it could be used to promote many beneficial outcomes as the profits could be used for the benefit of all. If it really takes off it could displace many millions, or even hundreds of millions, of people currently working in agriculture. As these new factories would be highly automated, they would only create a tiny fraction of the employment that farming currently does. Under the kind of Global Green New Deal outlined in this book this economic disruption would be liberating for people, whereas under market fundamentalism it would result in mass impoverishment. How we apply any technology is ultimately a political decision. Potentially this could be one of the key breakthroughs of the twenty-first century, given the right political and economic framework for it to be developed most beneficially.

Rewilding

The Knepp estate in Sussex has been extraordinarily successful in bringing back many species of plants, insects and birds to 3,500 acres of the Sussex countryside. Isabella Tree's book 'Wilding' is a wonderful heart warming account of their journey. Knepp provides a model and an inspiration on which we should build. Knepp now is primarily a project aimed at allowing nature to bounce back. It does still produce some food; mainly beef from their herd of longhorn cattle, but this food production is almost a by-product of their ecosystem restoration, rather than a central goal of the project. They run wildlife safaris and it is this eco-tourism that is the mainstay of their business. Also old farm buildings have been converted into offices and workshops, which generates another income stream. Now more people work at Knepp than in the days when it was an arable farm. Many people associate rewilding with removing people from the landscape. The truth is that people and wildlife can flourish side by side. This is an important point when considering larger scale rewilding.

Almost all of the proposed 39,000 model farm based land use projects would have some element of rewilding. Most of these projects would primarily be focused on food production, but quite a few would have rewilding as their core activity. Knepp has been very successful on 3,500 acres. Many people would like to see very much larger projects.

Wild East is a small new organisation promoting the idea that as many as possible of the landowners across the East Anglian region of England devote 20% of their land to rewilding. They hope to have farms, private gardens, schools, and other areas involved. Their model seeks to integrate rewilding into one of the most intensively farmed and settled areas of the UK. If all the vast network of hedges that exists across the farmed landscapes of lowland Britain was left to grow a bit taller, wider and wilder this would be a tremendous gain for wildlife. Their interconnected pattern would act as corridors allowing plants, animals and birds to move more easily from one area to another.

At the other end of the UK spectrum there is much interest in undertaking a lot of rewilding in sparsely populated northern Scotland. Alan Watson Featherstone started the Trees for Life organisation in 1993 with the aim of restoring the Caledonian Forest, the indigenous old temperate rainforest of highland Scotland. They have worked at Glen Affric and bought land at Dundreggan.[78] Various organizations have nature reserves in and around the Cairngorms National Park and elsewhere in Northern Scotland. The time is now right to consider a very large scale rewilding project, linking together all these initiatives and more into one unified project.

Jeremy Leggett, an early pioneer of solar power in the UK, founded Solar Century in 1998, building it up to become the UK's largest solar business before selling it in 2020. He moved to the Scottish Highlands with lots of exciting ideas for large scale re-wilding with strong social and economic benefits as a core goal. First he purchased 511 hectares of land at the Bunloit Estate, Inverness-shire, and soon after the 349 hectare Beldorney Estate in Banffshire. The Bunloit Estate is a rich mix of habitats already and will be the centre for research. The Beldorney Estate is mainly made up of overgrazed pastures and monocultural conifer plantations, with a historic castle surrounded by a small area of ecologically rich riverbank broadleaved woodland. The overgrazed pastures and conifer plantations provide an excellent area to treat as a blank canvas for rewilding. Jeremy Leggett has built up a team of researchers and rangers to help bring his wide ranging set of goals into reality.[79] These plans include setting up a mass ownership company called Highlands Rewilding Ltd as the vehicle to buy a lot more land across northern Scotland with the intention of creating new forms of employment based on creating a richer ecological habitat than currently exits. The scope for this is tremendous. Thousands or hundreds of thousands of people might invest financially and become stakeholders.[80]

My envisaged Green New Deal assumes vast sums coming from global taxation. In reality the kinds of projects I am advocating may be initiated more in this way, like Highlands Rewilding Ltd, as individual projects seeking investor stakeholders. This book portrays what could be done, and what may be done subsequently, but first pioneering projects like this are showing the way to go.

Much of northern Scotland, as indeed many parts of the UK, are over-grazed pastures, non-native conifer forests and grouse moors, all of which are relatively species-poor and essentially denuded landscapes. None of these land-use systems is that profitable or employs that many people. There are dozens of potential sectors that could be developed, from eco-tourism to renewable energy, better quality timber such as native oak as the basis for furniture making or for eco-building. Pockets of the most fertile land could be developed for market gardening, and greenhouses could be used to extend the growing season and the range of crops grown. Most of the land would be mainly for biodiversity gain and carbon sequestration, with all the other economic activity taking place against this background. The human population of the Highlands might increase just as the wildlife populations do; they do not need to be thought of as in competition.

Many species, like White-tailed eagles, ospreys and beavers are already breeding successfully in northern Scotland. Many more species could be introduced. The re-introduction of lynx, wolves and bears is controversial and may still be some way off, but I think if it is to happen anywhere in the UK this is the region to do it. I feel the benefits would vastly outweigh the dangers. Our proposed Global Trust could help fund and coordinate such a large-scale rewilding project, within which some model farms might be exploring the possibilities of combining innovative, sustainable and highly productive farming practices focused on food production in small areas within a context of very large scale landscape rewilding.

There are many ecosystems that have great potential to sequester carbon and are often little understood. Wet peat bogs covered with healthy sphagnum moss sequester carbon, and peat being almost 100% pure carbon, peat bogs are themselves tremendous carbon stores. Over the years peat has been dug up for fuel in domestic houses and in power stations, and to be sold as soil conditioner in garden centres. Peat bogs have been drained for agriculture or for tree planting. All these abuses of peat bogs need to stop. At last now some peat bogs are beginning to be restored across many parts of the UK. Much more needs to be done.

Tidal mudflats are another ecosystem that have often been regarded as unproductive when in actual fact they are productive and ecologically rich. Many such areas should be rewilded, but some could be developed in ways that are both highly productive and also support elements of rewilding. A few pages back I looked at the work of a new start-up organisation called Seawater Solutions, and their model seems adaptable to achieve an interesting combination of improved food production and increases in biodiversity ideally suited to reimaging tidal mudflats.

Rewilding and Forests

The forests of the world are an extraordinarily diverse collection of ecosystems, and everywhere they are currently under threat. Forests provide so many vital ecosystem services including carbon sequestration, regulating the water cycle, providing habitat for extraordinary biodiversity and livelihoods and home to many people. There are about 4.06 billion hectares of forest globally. 'Since 1990, it is estimated that 420 million hectares of forest have been lost through conversion to other land uses'[81]. Of the still existing forests many have been very badly damaged by the felling of ecologically rich old growth forest and replanting, often with monocultural forestry, which supports a tiny fraction of the biological diversity that the former old growth forests would have contained. The primary cause of forest loss has been for conversion to agriculture. The resulting agricultural land often has poor soils, and is then often poorly managed allowing soil erosion, declining crop yields and in some cases desertification.

The thinking behind my proposed Global Green New Deal, the Global Trust for People and Planet and the network of 39,000 model farms would be to preserve and even enhance all of our remaining forests globally. This would be achieved in several ways. The global network of model farms would seek to demonstrate how it is possible to grow a lot of food on a relatively small area of land. The Westland area in Holland is only 90.74 square kilometres or 9074 hectares, and includes a lot of built up areas. It does also contain a lot of extremely productive greenhouses, contributing to the astonishing fact that densely populated and tiny Holland is the world's second largest agricultural exporter. Most of our network of 39,000 model farms would include some very intensive systems of food production, often utilizing greenhouses. I have looked at this elsewhere in this chapter. I have described a range of other technologies that could mean that most of the world's food could be grown on a smaller area than is now the case and that much of this new food production may be in areas such as cities and deserts that currently produce very little food.

I mention these technologies in connection with forest preservation because it is pressure to expand agricultural land that has led to forest loss over many decades. This could be reversed if our network of model farms increased global food supply from a tiny percentage of the land area. That is the most important thing that can take pressure off our forests. The model farms would also have another key role in forest preservation. They would provide alternative and better paid employment with a much better and more secure lifestyle than currently can be achieved by the farmers, ranchers, poachers, miners and charcoal makers who currently contribute so much to the destruction of forests. Also some forest areas, especially the eastern Congo basin, but also areas in Papua New Guinea, The Philippines, Burma, Columbia, Peru and elsewhere, have seen civil war or insurgency issues resulting in much human suffering. Our network of model farms would also have a role in de-escalating such trouble spots. In Chapter Five we will explore how such a project might look in the context of a hot dry desert region such as the Middle East, and how the huge employment and training opportunities afforded by my proposed Global Green New Deal would seek to divert people from conflict toward co-operation. I will touch on how this might be applied in a rainforest region in that chapter.

The vision I have for the world's forests is inspired by my reading about many indigenous cultures' ways of living in productive harmony with their forest environments. Protecting and preserving forests has often in the past been thought of as keeping people out and thinking of them as 'wildernesses'. However there is a growing body of evidence that the indigenous peoples of the forests are their best custodians. Many of these indigenous peoples naturally now want the benefits of modern life. By combining the best of the indigenous peoples' stewardship of the forests with the best modern agroecological knowledge and the funding proposed in my Global Green New Deal, it ought to be possible to achieve the most amazing results for people and for planet.

The Spanish Conquistadors, as they spread through South America in the decades following 1492, noted how the indigenous people in Amazonia incorporated charcoal and organic matter into the naturally poor and acidic soils and made them extremely fertile. Tragically diseases and genocide brought collapse to many of the peoples of Latin America, including in Amazonia. The terra preta soils, or Amazonian dark earth soils, are a human made phenomenon, still fertile hundreds of years after they were made. Many species of plants were domesticated and grown on these soils and much if not all of the wider Amazon forest was in fact gardened with intent, with the most useful trees and plants deliberately selected and in some cases planted. The Brazil nut is a fine example.

There is so much to learn from this. We could imagine a network of model farms within the forests making small vegetable plots, utilizing biochar, compost and manures to build optimally productive soil.

Many of the world's best botanists are from rainforest cultures, able to identify thousands of species and know the many uses of them all, from foodstuffs to medicines, building materials to making into canoes. Brazil nuts are hard to grow outside the intact Amazonian rainforests, as they require pollination by a very particular kind of bee that only lives in healthy rainforest. Brazil nuts are extremely nutritious, full of protein and useful minerals and essential elements such as selenium. Some of our model farms could be centres of research, development and deployment with the aim of massively increasing global supply of Brazil nuts and other useful Amazonian forest products, while also preserving and enhancing those very same forest areas for biodiversity, for eco-tourism and for many other mutually compatible goals.

The world's forests are an extraordinary range of ecosystems, including many types of tropical rainforest, cloud forest, seasonal monsoon forests and drier and more open tropical woodlands merging into the partially wooded savannahs. Then there are the vast boreal forests, the deciduous woodlands of Europe and the vestigial temperate rainforests of Wales and Scotland. Many small pockets of forest still contain species of plant and of animal that are endemic just to that one location. We must strive to preserve them all. They are unique and deserving of protection and of serious financial investment, not primarily as resources to be exploited but in their own right and as symbols of humanity's capacity to manage the planet in ways that are intelligent and compassionate to all living things.

Although the general picture is of forest loss there are many examples of places where things are beginning to turn around, as I have just been exploring in the context of northern Scotland. Costa Rica is perhaps the best example. From the 1940s to the 1980s logging destroyed much of its ecologically rich rainforest. From 1987 it started to turn things around, by taxing fossil fuels and paying farmers to protect existing forest and to re-establish forest species on their land.[82] Cattle made way for a rich mix of productive vegetable gardens, fruit trees and forest trees. As eco-tourism increased it brought more money into the economy and into the rural communities, and helped local people see the economic benefit of having species rich forests. Gabon in West Africa now seems to be implementing some good policies with the intention of preserving and increasing its rainforest cover. I will explore some of the possibilities of how these positive trends could be further developed in the next chapter.

Rewilding Oceans

Oceans cover over 70% of the surface of the Earth, and about 97% of the Earth's water is in the oceans. Most of the 3% that is fresh water is in the form of ice at the poles and in glaciers, with less than 1% as liquid fresh water and a tiny fraction as water vapour in the atmosphere. Most of this water is in some way polluted and acidified; ecosystems are damaged and denuded. One of the major challenges of the twenty-first century will be to clear up this mess. The obvious first step is to stop the damage, and my proposed Global Green New Deal would seek to achieve this by clamping down on all that is damaging, from single use plastics to carbon emissions, overfishing and seabed damaging systems of fishing such as trawling. The second step is to create areas where ecosystems can successfully flourish. Many marine ecological hotspots are now marine reserves, but without success on the macro-level challenges, they are doomed to failure. So for example many initiatives seek to protect individual coral reefs and even to hand propagate small pieces of coral, and such work is doomed to failure if we do not stop ocean acidification and ocean warming. My proposed Global Green New Deal would focus on getting these macro trends reversed. If we can assume that action to be successful, then there is much that can be done to enhance biodiversity in our oceans.

Marine conservation zones, such as the waters around Lundy, have been in existence since the early 1970s. Many of the newer protected areas are much bigger.[84] There are many different types of protected areas. Some are no-take zones, intended to be left for nature to flourish and where human activities are reduced to virtually nothing. As we have seen with rewilding in general the principles can be applied to heavily populated areas such as Wild East in the UK's East Anglia region, and the recent achievement of making London a National Park City certainly expands the concept of what a national park can be. I want similarly to expand the concept of marine protection.

In Talamone in southern Tuscany, Italy, an interesting part art installation part marine conservation project is happening as wonderful Carrara marble sculptures are lowered on to the seafloor.[85] This has several functions. It should stop illegal trawling as the blocks of marble would damage the fishermen's nets. It also puts Talamone on the map and opens up a potential underwater tourist attraction as people can scuba dive to see the sculptures and to see how marine life colonizes them over time. The protection that the sculptures give to fish should allow fish stocks to replenish, helping fishermen in the longer term.

Greenpeace have a campaign of dropping large boulders on to the sea floor to discourage trawlers from illegally fishing in a marine protected area off the Sussex coast. Obviously trawling and other destructive fishing practices need to be banned. What Greenpeace are doing is in effect starting to build an artificial reef.

Building artificial reefs to enhance marine ecosystems is possible, but must be done with the right materials and in the right places. In the past it has been used as a ploy to get rid of rubbish such as old car tyres, which over time release a horrid cocktail of pollutants. Obviously such practices must stop. But done well, artificial reefs do have possible advantages, and are yet another area justifying considerable research, development and deployment funding.

Tidal lagoons, as proposed for Swansea Bay and other locations in the UK and globally, have great potential to deliver multiple benefits, including generating electricity, aquaculture production, leisure pursuits and the creation of more ecologically rich and diverse habitats. So far none has been built, but I have long been in favour. I think the benefits outweigh the costs. Offshore windfarms have been developed in the North Sea and elsewhere, and very many more will be built over the coming decades, and as floating turbines are developed, the global geographical spread of offshore windfarms will rapidly increase. Tidal lagoons and offshore windfarms will change our coasts and oceans. They could be the basis for whole networks of artificial reefs. The underwater sections of wind turbines already act as artificial reefs and are colonized by seaweeds, mussels and a range of other species, providing food and habitat for fish and seals. By hanging ropes or chains between the turbine towers this reef effect could be much enhanced.

With sufficient research and development funding, and with the right rules and institutional frameworks, this mix of renewable energy infrastructure and reef construction has the potential to add to biodiversity while also creating social and economic benefits. Merging the concepts of rewilding and economic development may sound like a terrible compromise, but I think it will become important in many places. If London can become a National Park City, and seek to develop many more green spaces for wildlife to flourish, people to thrive and pollution to be reduced, while also providing homes and livelihoods to many more people, then surely we can create a similarly wide range of apparently contradictory benefits on certain parts of our coasts and oceans.

On the David Attenborough Green Planet programme, first broadcast on 16th January 2022, he cited the extraordinary speed with which healthy seagrass meadows can absorb carbon dioxide from the atmosphere, in some instances up to thirty times faster than a tropical rainforest. Maybe it is time to invest vast sums of money into trying to rapidly expand seagrass ecosystems. Many shallow seas have been damaged by dredging and trawling and would be suitable for such large scale restoration and new habitat creation.

Rewilding Rivers

Nearly all the rivers and streams in the world are to some extent polluted, but the degree of pollution, and the types of pollutants vary enormously. Concern about local level pollution and taking action to combat it goes back a long time. Take the river Thames as an example. As London grew in the eighteenth and nineteenth centuries, increasing quantities of sewage and other pollutants entered the river, causing cholera epidemics, and then with the hot weather of the summer of

1858, the famous 'Great Stink'.[86] That prompted government action and Bazalgette's sewage system was built. The important lesson here is that it takes governmental action to clear up pollution. In the 1950s the River Thames was declared dead. It was polluted to such an extent that no fish could live in it. Many rivers in other industrial areas of Europe and North America were similarly dead. During the 1960s and 70s legislation was enacted in many of these places, some agricultural chemicals were banned and industrial outfall pipes monitored, so consequently rivers improved, and 125 species of fish were recorded in the River Thames.[87] In recent years there have been government cutbacks and a decrease in monitoring by the Environment Agency. Inevitably pollution has increased again.

Where I live the River Wye is in trouble.[88] Here the main pollution comes from agricultural run-off. Phosphates from farm animals, especially intensive poultry units, are one of the main factors contributing to the decline in river quality. There are many other factors, including a range of toxic chemicals used on farms, spills from sewage works and septic tanks, soil erosion and litter (especially plastic). These are the main issues locally, but there are other sources like the oil and particulate run-off from roads and spot pollution incidents from industrial processes and spillages. Fish stocks, and the bird life that is dependent on them, have declined dramatically. Algal blooms are appearing further upstream and earlier and earlier in the year, robbing the river of oxygen and resulting in mass fish deaths.

The proposals in my Global Green New Deal would seek to clean up the Wye, the Thames and all the world's rivers. Many chemicals would be banned, others taxed, intensive poultry broiler houses would be banned and the numbers of farm livestock reduced and their grazing systems closely monitored and kept back from watercourses. The use of plastic would be reduced and infrastructure such as drains and sewage works improved. Local government would have vastly increased funds and responsibilities to monitor pollution and to take strong enforcement measures. Reducing all forms of pollution is possible, but it requires consistent political determination, commitment and funding over many years.

There are of course plenty of other issues affecting rivers. Hydro-electric dams inevitably cause huge changes to rivers. River valleys vary enormously. Some hydro-electric systems have been built where the ecological damage has been terrible and the economic benefits minimal. In the chapter on energy we noted how the Balbina dam in the Amazon rainforest is one of the worst and Kvilldal pumped storage hydro system in the mountains of Norway one of the best. Over the coming years it would be good to have more hydro schemes like Kvilldal and less like Balbina. The design and location of new hydro is important. Maybe it is time to remove some of the worst dams and let places like Balbina revert to rainforest and the Uatuma River to once again be a wild and free river.

In many places rivers have been dredged and straightened with the aim of trying to transport the water to the sea as quickly as possible. This, in combination with the decreased level of soil permeability and the loss of woodland on the watersheds has contributed to many more flooding incidents. It is now increasingly being understood that a better way to manage river flows is to help the water percolate into the soil by increased tree cover, especially along the watersheds, with more hedges across the angle of slope on fields to reduce run-off, and by slowing down the flow of water, especially in the upper catchment to protect downstream communities from flooding. Beavers, with their leaky dams, have a very useful role to play in this ecosystem enhancement, improving biodiversity while also contributing to enhanced flood control.

In some places rivers and streams have been put in culverts and sometimes used as sewers. As with hydro schemes like Balbina, or the dredging and straightening of rivers, maybe it is time for some of them to be liberated and returned to something approximating a healthy wild state. In all these situations careful assessment needs to be made of what is right in each individual situation. In some small ways rivers are being rewilded, and this will prove useful as we seek to do very much more of this. There is much to learn.

Ecosystem Restoration

Over the coming decades it is probable that very large scale ecosystem restoration will take place. In any ecosystem rewilding can go hand in hand with more sustainable forms of prosperity. The key is always first to identify the most damaging practices and reduce or stop them entirely. This then leaves space for more creative and ecologically literate forms of resource management. We have seen this in the rainforests of Costa Rica, and in some of the best nature reserves, marine reserves and other situations. Now it is time to take this process to the global scale. Some places will be tremendously challenging.

The Alberta tar sands region will be polluted for many decades to come. Areas where fracking has been undertaken will have a toxic cocktail of carcinogenic chemicals in the ground, always at risk of leaching into aquifers and threatening human health. Decommissioning the Sellafield nuclear site will take at least a century and cost a fortune. It would have been better if none of these projects had ever been developed. Now it is the task for our species to clear up this mess. Some places may not be fit for human habitation again for tens of thousands of years. The International Criminal Court should make ecocide a crime. The corporations and governments responsible should be made to pay for the clean-up. Many trillions of dollars will be needed. Many corporations should be closed down and all of their assets seized. Many politicians and business leaders should be serving long sentences in jail.

There are many old mining and former industrial sites that are now nature reserves. Forests have now colonized the whole Chernobyl area. Nature can recover, but watercourses and aquifers will need careful monitoring and expensive remedial action will often be required for a very long time into the future, even if we stop doing the damage today.

We have looked briefly at ecosystem restoration following industrial pollution. There is another aspect of ecosystem restoration that will be on an even bigger scale, but where the pernicious problems of toxicity are not the problem. Many of the world's hot dry deserts have extended their range as people have cut down trees and cleared land for agriculture, only for it to turn to desert. This process has been going on for thousands of years, but has increased as populations have grown and more trees are cut down. Global heating is making the situation very much worse. Reversing desertification provides extraordinary opportunities if done well. Much of this book is about new ways the deserts of the world might be made productive with solar power, desalination and a range of newer technologies and land-use systems, and these new land-use systems should be powerfully carbon negative so helping us all avoid the worst of the unfolding climate catastrophe. Many areas of desert could be turned into forest gardens, mixing trees, agrivoltaics, fruit and vegetable growing, and some livestock, always with one of the aims being soil creation and carbon sequestration.

Global scale ecosystem restoration is of fundamental importance. It will require very strong co-operation between local communities, countries and the global community. Rewilding and new and ecologically literate systems of economic development must replace our current most damaging industries and economies. In the next chapter I try to assemble the pieces of a geographical jigsaw to show how this might come about.

Reconnecting People with the Land

Over the last sixty to seventy years people have become ever more disconnected from the land and from each other. Many people know very little about how their food is grown, or how to cook healthy food. Often people feel lonely, isolated and depressed. In his book 'Lost Connections' Johann Hari explores the causes and unexpected solutions to anxiety and depression. Connecting people to nature and to each other in communities with a sense of solidarity and shared physical activity are all part of his recommendations for restoring mental and physical health.

Our network of model farms would play a critical role in helping with the global epidemic of anxiety and depression, obesity and poor nutrition, and many other aspects of improving human and planetary health. Our model farms would seek to grow and sell as much food as possible directly to the people who are to eat it, and as little as possible to wholesalers and supermarkets. The farms would seek to build very strong connections with their local communities. Food could be delivered directly to customers by electric cargo bikes, as most people would have one of the model farms within a fifteen minute cycle ride.

Most of the network of model farms would have thousands of people coming to help with the harvests, to camp and share food and to share festivities and educational projects on the farms. Many of the farms would have kitchens and restaurants and run classes in cooking. They would all welcome schools visiting throughout the year, learning about where their food comes from and about nature. Each of the model farms would have some element of rewilding and of helping people reconnect with nature. Many of the model farms would to some extent have aspects of care farming, where people with various health or disability issues could be helped to find healthy, sociable and purposeful activity on the farm.

Our network of model farms would be distributed globally within communities where the challenges and the opportunities vary enormously. In the next chapter I shall explore some of the possibilities of what we could do if some kind of Global Green New Deal is enacted and there are massive funds to invest. In the remainder of this chapter I just want to explore half a dozen or so farming projects that have inspired me.

The Fold is a very interesting project at Bransford in Worcestershire. At its heart are an organic market garden and a small care farm.[89] I remember being shown around a few years ago by Lucy, who then managed the care farm. I was really impressed with what they were doing. It was such a welcoming place. They sell veg boxes and supply local restaurants, and have their own on-site cafe. They host all manner of events from camping conferences to music gigs. They have developed nature trails around the farm and old buildings have been sympathetically renovated to improve their energy efficiency, beauty and usefulness. Artists' workshops and therapists' rooms occupy old farm buildings. Many hundreds, probably thousands, of people come to the farm every year.

Bore Place in Kent has many similarities with The Fold in that they welcome thousands of people each year to participate in the life of the farm in lots of ways. Bore Place was purchased by Neil and Jenifer Wates in 1976, and in 1977 they set up the Commonwork Trust. For the last forty-five years they have been gradually adding more elements to the mix. I visited in the very early days, probably 1980 or '81. I was impressed then and have watched them develop since. They have an organic farm with a dairy herd and make organic cheese. This is combined with an organic market garden and care farm. They have renovated many old farm buildings, and in the process developed their own brickworks, and they have developed on-farm renewables, including an anaerobic digester, biomass boilers, wind turbine, solar hot water and photovoltaics. They work with Kent County Council to provide residential breaks and day workshops to people with special needs. Like The Fold Bore Place is a farm teeming with richness and diversity, of crops, wildlife and people. Bore Place is on a somewhat larger scale than The Fold. The next project I want to look at is on an altogether bigger scale, and in a very different climatic and cultural environment, but it too incorporates many of the same rich mix of activities.

In 1977 Dr Ibrahim Abouleish and his wife Gudrun purchased 70 hectares of desert land outside Cairo in Egypt and started the Sekem project. This has grown mightily over the years. They developed methods of biodynamic farming growing a wide variety of medicinal herbs, and culinary herbs, fruit and vegetables. They developed organic ways to grow cotton that used very much less water, and no chemicals, unlike most cotton production. They started care farming, then started running their own school, medical centre, library, and eventually their own university. They opened shops in Egypt and exported produce to Europe and to the USA. They have worked on comprehensive poverty reduction programmes in villages and their farming enterprise has grown considerably. They have won lots of awards for their pioneering work and have spun out numerous new start-up businesses and they partner with many international organisations.[90] They have pioneered an ecologically restorative sewage disposal system. The original 70 acres of desert is now super productive organically rich farmland with plenty of tree cover. Also their organic farming systems allow wildlife to prosper in a way that the arid desert could not. It provides a template for some of the projects I have in mind for some of our envisaged network of model farms.

These three projects, The Fold, Bore Place and Sekem all exhibit a welcoming approach to people. They have all achieved a lot in terms of good quality organic production, and are all good places for nature to heal, but what is so impressive is that they have all achieved this by bringing many more people on to the land in welcoming and caring ways.

By contrast so much of our farmed landscape seems to be in the hands of farmers who don't want people on their land. The land needs people to flourish and many people need contact with the land in order to thrive. The farms that most vigorously try to keep people away are often those farms using excessive chemicals and machinery, and often also the most inhumane systems of livestock management. They tend to be unhappy places.

Earlier I looked at White Oak Pastures, which the Harris family has farmed for four generations. In the post war years they went down the chemically intensive path, including hormone implants. In 1995 Will Harris changed course in quite a radical direction, pioneering an organic system of pasture fed multi-species rotational grazing, with on-site butchery and direct marketing. Will Harris has said that although the farm was profitable under the old chemical model, he enjoyed it less and less each year. That was one of his motivations to make the transition. The farm has since developed all manner of additional features, such as its own building crew, lodges to rent, and educational workshops. Before 1995 the farm employed just a couple of people, now it employs over one hundred and fifty-five full time staff and hosts many visitors.[91] It has become a happier place. Unlike The Fold, Bore Place or Sekem they are not a care farm, but like them they have successfully brought people back into the farmed landscape.

There is another farm that I want to mention. Angus Davison started

Haygrove in 1987 in Herefordshire, initially growing just strawberries. Now just thirty-four years later they have farms and market soft fruit all over the world, and have many thousands of employees. Their main five crops now are strawberries, raspberries, blueberries, blackberries and cherries, some of which they grow organically. They pioneered the use of multi-bay polytunnels, which many people dislike, but which have made growing berries and cherries commercially viable again. I would think that in a few years most of their polytunnels will be replaced by the kind of clear photovoltaic panels now being trialled in Holland. Their farming system would be ideal to convert into agrivoltaics, producing both electricity and fruit. No doubt some people would object, as they always do to anything that is not exactly how it was in the past.

I really like Haygrove for a number of reasons but the one I want to focus on is the opportunities they offer to fruit pickers. Many farms treat casual fruit pickers pretty badly. In 2007 Haygrove introduced their Bright Futures social uplift programme, offering five years of training so that some of their fruit pickers could become partners within the business. As they have expanded globally much of this has been led by people who began as casual fruit pickers. In 2013 Haygrove started partnering with people establishing local community gardens, a kind of care farming, the first one being in Ross-on-Wye.[92] Like White Oak Pastures they employ a lot of people directly and now they are developing this care farming aspect. They too are bringing people back onto the land.

Over a couple of years in 1980 and 1981 I visited and worked in a number of communes and intentional communities trying to grow as much of their own food as possible. The Wheatstone commune in Herefordshire was back then perhaps the most radical in that they strictly limited the products they allowed themselves to buy. So they did not eat rice or oranges or most other imported foods. Instead their diet was almost entirely made up of the fruit, vegetables, meat and dairy that they raised themselves. They did buy in oats, wheat flour and tea, but not a lot else. It was a very interesting experiment and somewhere to stay for a few weeks. Canon Frome Court was more an intentional

community, divided up into family units, with about fifty people of all
ages collectively working a forty acre mixed organic farm, growing their
own grain, as well as fruit and vegetables, meat and dairy. [93] I particularly
liked working in their two acre walled organic vegetable garden. I
still have friends there, and have watched this land being worked for
forty odd years. It must be one of the most productive couple of acres
anywhere. It provides yet another model of re-peopling the land in
ways that are ecologically sustainable and productive. Such co-housing
and community level farming projects could become a major part of
providing housing and food for millions of people.

There is a long history of urban populations wanting to return to
the countryside, and often to live and farm in co-operative ways. Denis
Hardy wrote a book 'Alternative communities in nineteenth century
England'[94] and he led a workshop on this at Canon Frome Court back
in the early 1980s. This spirit is very much alive today. There are many
people planning and developing all manner of eco-villages, agri-villages
and the like. The whole community supported agriculture movement
links in to all of this, reconnecting people with where their food
comes from.

My envisaged network of model farms takes inspiration from this
whole history of intentional communities, eco-villages, community
supported farming and from these half dozen or so farming projects,
and many other ideas, projects and technologies. It would seek to take
this re-peopling of the landscape a whole lot further. Let us envisage
investing trillions of dollars setting up a global network of model farms
creating hundreds of millions of new jobs and livelihoods. The goal
would always be to simultaneously help heal as many of the world's
problems as possible.

Chapter Five

System Change & Regional Geographies

System Change & Regional Geographies: An Introduction

In previous chapters I have looked at system change across many disciplines, from governance to food and farming. In this chapter I will take a regional geography approach. The kinds of actions that I have been proposing would change every part of the surface of the Earth, but these changes would affect different places in very different ways. This chapter gives us the opportunity to see how the changes mentioned previously might play out in specific locations around the world.

We know that to create the desired system change we have to imagine the future without many of the things we currently take for granted. Less well understood, but of even greater importance, is the ability to imagine a future where many things that currently seem unachievable or utopian, or little known or understood, become established daily reality.

So can we imagine a world without any fossil-fuelled internal combustion engines, and where the private ownership of cars is extremely rare, and for most cars to be collectively owned by car sharing clubs, and of course all to be either battery electric or hydrogen fuel cell? We will look at how this might play out in big old cities like London, small cities like Hereford, or across the rural Welsh Marches. As air travel is so very polluting and so technically difficult to switch to cleaner technologies, can we imagine a world with perhaps 90% or 99% less air travel? How would such a scarce resource best be allocated in a manner that was socially just? How might new and slower forms of less polluting international travel change how people moved between places and how might this fit in with changed work practices?

Can we go much further and imagine a world without nation states, armies or wars? In this chapter we shall explore some of these issues across various locations, including in some of the world's most troubled regions, such as the Middle East. To create system change we do need to discuss ideas about what a very different system and a very different future might look like and how we might get there.

Epochal Shift & Global Demographics

In earlier sections of this book we have looked at the epochal shift, from 'The Fossil Fuel Age' to 'The Solar Age', and how a range of possibilities and opportunities are opening up in the sunniest countries of the world. Solar power is rapidly becoming the cheapest form of energy, and historically industry and people have always moved to where energy is cheapest. New cities may well emerge in what appear now to be some of the most unlikely places, in the hot dry deserts. In this chapter I will drill down into this and look at some of these solar technologies and how they may be the linchpin of whole new patterns of economic and social development. Deserts also provide an opportunity for massive photosynthesis driven carbon sequestration which could be done in many ways, some with great potential for food production, some for a massive gain in biodiversity and some a combination of both.

Some of 'The Fossil Fuel Age's' most iconic cities are in deserts, exemplified by Dubai, but those of 'The Solar Age' may appear in many more places and be characterised by very different technologies and values. As these new solar cities emerge from very different cultures they will no doubt have many differences, but they will also have much in common, united in their need for water and the technological opportunities they have to develop their solar resource. Co-operation between them will be of immense mutual benefit.

If one looks at an atlas of the world's solar resource the hot dry deserts leap out in vibrant oranges, reds and, in very most intensely sunny locations, as deep maroon.[95] Already solar power stations are being built in most of these desert regions, but at present they represent a tiny fraction of humanity's overall energy production. This will increase, and my job here is to try and sketch out how this may be done in the most beneficial way possible.

Much of the spending in my proposed Global Green New Deal would be focused upon these Sunbelt regions. In Chapter Two I looked at how we might raise and invest huge sums of money. There already exist a number of amazing technologies and projects that are emerging in the world's hot dry deserts. More are being developed all the time. The purpose of the proposed Global Green New Deal investment would be to multiply these kinds of projects and to add more and more elements to them to simultaneously solve our whole basket of problems. So let us examine our basket of potential solutions and see how they could best be put together in specific locations.

The Middle East Reimagined

How might the Middle East look if we applied the proposed Global Green New Deal? It is a region beset by multiple problems. It is also a region ideally suited to becoming an epicentre of the new ideas and technologies of 'The Solar Age'. Could these ideas and technologies be put to use to solve the regions many problems?

Some of the huge problems we would be seeking to mitigate or solve are:-

1. Israel-Palestine: decades of conflicts, settler-colonial relations, tensions
2. Ongoing extreme political instability: eg in Syria, Iraq, Yemen, etc.
3. High numbers of refugees in Jordan, Palestine, Turkey etc.
4. Poverty and lack of basic services and opportunities for many (eg Egypt)
5. Extreme inequality: excessive wealth and absolute poverty are common
6. High unemployment: increases poverty and wastes vast human potential
7. Oil dependency and very high carbon footprints (Saudi Arabia, UAE etc.)
8. Water insecurity: over-abstraction, falling water tables

9. Food insecurity: oil states have unsustainably relied on importing foods

10. Biodiversity is under threat here as elsewhere

The proposed Global Green New Deal would have a transformative effect across the region, as it would everywhere. By taxing wealth and extractive resource use, with these taxes mainly being set at the global level and invested at the community level, national governments would see their powers decline. Large corporations would also be constrained by more social and environmental legislation set at the global level, and excessive profiteering would be limited by heavier taxes. Stock markets would also be more heavily regulated and taxed, and their dominance over policy would be reduced. All this is necessary to allow political space and economic resources for our global network of local communities to get on with the many things that need to be done.

Currently many of the governments of the Middle East spend a high proportion of their revenue on 'defence', while failing to provide basic health, education or economic opportunities to their people. This of course exacerbates political instability. The global armaments industries need to be defunded, diminished and regulated out of existence. The shift of resources away from governments and corporations and towards specific projects to promote peace, justice and economic opportunities embedded within local communities and backed up by a massively empowered and democratized United Nations, would help in this process of the gradual disarmament of the entire region.

Is it possible to envisage the Middle East becoming a region of peace and co-operation? In the current political climate it is hard to imagine such a thing, just as during the Second World War it was hard to imagine peaceful co-operation becoming the norm for Europe. The three quarters of a century since the end of the Second World War have been the most peaceful for a thousand years of European history. The European Union was founded in large part to promote peaceful co-operation, and in this it has been a very powerful influence for good. As the EU had its foundations in the European Coal and Steel Community, might peace in the Middle East be founded upon their collective development of a solar powered and ecologically rejuvenated economy? As with the European Union a Union of the Middle East would expand beyond these initial areas of industrial and land based co-operation into all aspects of life. Other perhaps overlapping unions would emerge, just as many in Africa see the African Union building co-operation across that continent, inspired in part by the peace and prosperity that Europe has achieved through co-operation.

Attempts at bringing peace to Israel's conflicts with its neighbours and with the Palestinians have often been couched in terms of talking of a 'Two States Solution' or a 'Single State Solution'. Here I want to float the idea of a 'No State Solution'. My concept of a Global Green New Deal is built around the prospect of political power and financial resources shifting up to the global and down to the community level, and consequently the role of states and of corporations would inevitably diminish. This might be invaluable in the search for peace in the Middle East. Establishing innovative projects which directly seek to solve multiple problems and where people from many backgrounds of race, religion and nationality come together in shared endeavour seems to me to be the best path toward peace, prosperity and a better future. It is envisaged that these projects will employ many millions of people across the region, which will have impacts way beyond the lives of the individuals directly working on these projects.

In the Chapter on Energy we noted that some regions will remain net energy importers while others are net energy exporters. For example Germany will be a net energy importer, being heavily populated and industrialized, with relatively poor wind and solar resources to draw upon. The Middle East as a region has the potential to generate more than enough energy for all its own requirements and still have massive potential to export. Solar power will be by far the largest part of their new energy economy. The solar resource of the Middle East can be used to achieve multiple objectives.

Solar technologies are rapidly evolving. Many of the leading academics and technology companies are based in Europe, but the Middle East is developing new research, development and deployment projects, often in co-operation with a network of global players. Solar power is growing anyway, but not as fast as it needs to. So for example the Masdar Institute aims to increase the renewable energy share of the total energy mix of the UAE from 25% to 50% by 2050.[96] Achieving 100% by 2035 would be a very much more ambitious and exciting goal, and in achieving it many benefits could be realized. Solar power can easily provide all the electricity the region needs, and also the heating and cooling requirements of buildings and industry. Solar powered desalination will become a vital sector of the economy. Land and sea transport will also be based on solar generated electricity and hydrogen, and in time no doubt air transport will follow. Until it does, air transport will need to be rationed.

The Middle East as a Mosaic of Communities

If we take 'The Middle East' to include: Egypt, Iran, Iraq, Saudi Arabia, Yemen, Syria, Jordan, United Arab Emirates, Israel, Lebanon, Libya, Oman, Palestine, Kuwait, Qatar and Bahrain, the total population of the region is very roughly 400 million people, with Egypt the most populous and Bahrain the least. Instead of thinking about this as a region divided into sixteen very different countries, locked in age old disputes with each other, let us think of them as an interlinked network of communities. So with our nominal average community size of 200,000 people, these 400 million people could be portrayed as being 2,000 of our average sized communities, rather than sixteen countries. The very concept of The Middle East is both disputed and vague. Some definitions extend the region further than others, and anyway the term was only coined by the British in the early twentieth century. The point I want to make here is that thinking beyond the divisions of nation states can be helpful in allowing more solution-focused ideas to take shape.

In earlier chapters I have looked at how we might raise many trillions of pounds, dollars or Euros, and how we might invest this money. We envisaged a budget of perhaps half global GDP, which would mean half of $88 trillion, so an annual budget of $44 trillion. If we kept $5 trillion of this for global level investment in agencies like the United Nations and the World Health Organization, and allow nothing for the national level, that would leave $39 trillion to be spent at the local level, shared between our envisaged 39,000 local communities with populations averaging 200,000 people. This would mean that each community would have one billion dollars to invest annually. Even if it was decided to put in only 10% of this figure that would still mean each community would have $100 million annually to invest. So let us conjecture how some of our communities, each working through the proposed Trust for People and Planet, might best co-operate to simultaneously mitigate or solve all the ten key challenges listed on page 35. So we would be seeking to develop projects that would simultaneously lessen conflict and poverty, promote human rights, cut carbon emissions, enhance water and food security and allow biodiversity to flourish.

There are lots of amazing initiatives already started in the region and also projects and developments from outside the region which could be used as templates for initiating new projects within it. Our media in the UK and elsewhere are very poor in reporting peaceful, productive or innovative things as they are deemed not to be newsworthy. The next section of this chapter describes a number of these excellent initiatives and speculates about how they could be expanded and replicated. Some of the projects I am about to describe are only at the ideas stage, others are small and recent, others are huge and some many hundreds of years old already. The one thing they all have in common is that they provide us with models of how a better future might be achieved. Think of them each as a small piece of a complex jigsaw puzzle, and if we arrange them in the best way a picture will emerge. That picture is of a very different future: a better future.

Soil and Water Conservation

We could start building our picture of that better future by examining some of the pieces of the jigsaw that illuminate how ecosystems could be restored and food and water security enhanced. There are some heartening examples of successful projects already established and some great ideas which could be used to massively expand on what has already been achieved. Climate change is exacerbating desertification and fuelling conflict and migration. By investing in projects that combat desertification, food and water security can be improved, peaceful co-operation established, and at the same time carbon can be sequestered. I will first look at three examples from outside the area, but with crucial lessons for The Middle East. I will then look at some even more ambitious ideas and projects and speculate how some of the key technologies we have already looked at in previous chapters could be developed in tandem with these land-use changes so as to have a transformative role across the region and across the world.

Our first example, and one that I particularly like, is what has been achieved along the Wadi El Ku in Sudan's North Darfur province.[97] Rainfall, already sparse and erratic, had become an even more precarious resource due to climate change, and this had led to conflict. As the then UN secretary general Ban Ki-Moon said of the conflict in 2007 'Amid the diverse social and political causes, it began as an ecological crisis, arising at least in part from climate change.' In Darfur, as across the Sahel and parts of The Middle East, the conflict was largely between herders and pastoralist communities fighting over scarce water. With help from outside agencies the rival communities were brought together to help construct a series of weirs all along the Wadi El Ku. These weirs act to hold back the seasonal rains and allow the water to percolate deep into the soil, preventing downstream flash flooding and more importantly helping to make the Wadi very much more productive. One example is that by the Sail Gedaim weir 4,000 farmers can now make a living whereas before the weir was built only 150 farmers could be sustained from this same area.

A much greater range of crops can now be grown, and grown more reliably, thanks to the weirs. Millet, sorghum, lentils, cucumbers, okra, melons and importantly fruit trees like lemons and grapefruit are now all growing well. The fruit trees help water penetrate into the ground and also provide vital shade, keeping the soil cooler and more fertile. This list of crops could of course be expanded massively, and as Ali

Mohammed, a local farmer, said 'You give me the seed, and I will test it.' Fodder crops like alfalfa could be grown to help the pastoralists, high protein foods like peanuts and cashew nuts could be grown and larger fruit and nut trees could be grown, providing vital shade and a wide range of food, timber and other products. As the communities worked together in shared endeavour building the weirs a powerful healing influence was exerted. Rival communities that had recently been killing each other came together and even started inviting former enemies to share in wedding celebrations. Across the Middle East there are many thousands of such wadis that could beneficially have weirs, terracing and other soil and water conservation measures built in them. If, for example, Palestinians and Israelis were working together for mutual benefit, more peaceful and socially productive relationships might be established.

Another larger scale and longer term example of successful soil and water conservation is what the Kamba people from the area around Machakos in Kenya have achieved. In the 1930s the area was suffering from overgrazing, soil erosion and slowly turning into desert. During the Second World War Kamba soldiers serving the British were stationed in India where they saw how terracing hillsides and developing other soil and water conservation methods had improved the productivity of the land. After the war they returned to the hilly country around Machakos and began a process that over the following decades developed into a phenomenon that came to be known as the Machakos Miracle. They shifted their farming practices from being predominantly pastoralists to become predominantly horticulturalists. They planted trees on the hills and watersheds, built weirs across the gullies and streams and terraced the hillsides so that 70% of the entire area was terraced. All these measures allowed the rainfall to percolate into the soil rather than rush off in flash floods, and so soil erosion could be reversed and fertility built up. The area now supports five times the population it did in the 1930s and provides them all with a better quality of life. Of course many more improvements could be made with sufficient funding, but what the Kamba people did with very little external help is truly inspirational. [98]

The huge Loess Plateau in northern China is a very much larger example of efforts to combat soil erosion. The plateau is 640,000 square kilometres of the most fertile but easily eroded soil. Its fertility had been important in Chinese history, but a combination of deforestation, overgrazing and inappropriate cultivation of sloping land had led to massive soil erosion and the fertile ecosystem was gradually turning to desert. In 1994 efforts, and resources, were put into reversing the situation. The basic aim was similar to the examples of the Wadi El Ku and Machakos but on a very much larger scale. Trees were planted on hilltops, along watersheds and on some of the sloping fields. Other slopes were terraced, and gullies were blocked with weirs and reservoirs. More fruit and vegetables were grown on the terraces and in greenhouses, some specializing in mushrooms and other high value products, which helped lift many people out of poverty. There remains much work to be done but what has been achieved here in China, as in our examples from Sudan and Kenya, shows how desertification can be reversed.

Thinking Bigger: Ecosystem Restoration

The history of most of the world's deserts is that they expand and contract, driven partly by changing land use patterns and partly by changing climates. The Sahara Desert has been expanding southwards into the Sahel for a long time. Many people have had ideas, and undertaken interesting and exciting work to reverse the desertification of the Sahel and other regions of the world for a very long time, probably dating back millennia.

I remember talking to my grandparents in the 1960s about the work of Richard St Barbe Baker and the organisation he founded in 1924 called 'Men of the Trees' (which later became the International Tree Foundation) and later being inspired by the great Wangari Maathai and her Green Belt Movement in Kenya, and also the charity Tree Aid. The Dutch geographer Chris Reij reported on amazing examples of farmer led innovation in the techniques of re-greening taking place in Niger and

Burkina Faso, and elsewhere. Now I follow on Twitter dozens of young climate activists from Africa such as the remarkable Patricia Kombo who started the Pa Tree Initiative. Many of these people founded tree nurseries and all advocated the advantages of planting lots more trees.

An idea to plant a belt of linear woodland across the continent of Africa to limit the southward march of the Sahara has been around a long while, and it is really heartening to see how this has developed into one of the most exciting and large scale projects in the world today. The plan is to plant a belt of trees 8,000 kms long, from the Gambia and Senegal in the west to Eritrea and Djibouti in the east. The African Union and governments across the continent are putting their enthusiasm and support into the project, and with help from outside agencies, have planted many millions of trees and restored tens of millions of hectares of degraded land.[99] The Great Green Wall organisation claims that they are about 15% of the way toward completion of this epic task.[100]

All these examples, from Wadi El Ku, Machakos, and the Loess Plateau to the Sahel's Great Green Wall, use soil and water conservation techniques to best utilize sparse and erratic seasonal rainfall. They are not focusing on creating new sources of water, and they are working in the semi arid zones rather than the true hot deserts. Lots of other ideas are swirling around which seek to amplify the possibilities of greening the deserts utilizing a range of innovative technologies. With solar powered desalination, using seawater or saltwater greenhouses or directly using solar power to drive techniques such as reverse osmosis desalination equipment, seawater, or brackish groundwater, can be transformed to fresh water. These newer technologies, coupled with some well established projects, could form the basis of the transformation of The Middle East, opening up both semi arid zones and true deserts to new forms of human settlement.

The Dutch engineer Ties van der Hoeven, inspired by a film about what had been achieved in the Loess Plateau, linked up with a number of other people fascinated by the possibilities of taking these ideas forward and formed an organisation called the Weather Makers. They have a plan to turn the Northern Sinai Peninsula green.[101] Lake

Bardawil, a lagoon on Sinai's Mediterranean coast, has been silting up for years and the once rich fishing grounds are deteriorating. They aim to rehabilitate Lake Bardawil by dredging it and use the clayish silts to help build weirs, terraces and bunds in the whole catchment of wadis that flow north into the Mediterranean. They hope to transform the eroded landscapes of the Sinai in ways very like what has been achieved in Machakos and the Loess Plateau. To harvest additional water they plan to use fog nets, a simple low cost technique that has been used in other areas of desert. With additional vegetation the expectation is that the increased evaporation will lead to increased rainfall.

Solar powered desalination is beginning to emerge as what will probably become a very important technology for the hot and arid lands of the world, wherever salty or brackish water is available to desalinate. The charity Give Power has developed a simple solar photovoltaic panel covered barn housing a reverse osmosis desalination unit capable of making sufficient fresh water for 35,000 people.[102] They have built a couple of such units in Kenya. What if we built a few thousand such units across the Middle East, sufficient to provide water for many millions of people?

In the previous chapter we looked at seawater and saltwater greenhouses, where salty seawater is desalinated using the power of the sun and wind, and the resulting fresh water used to irrigate and cool efficient and highly productive greenhouse based horticulture in arid desert lands. We looked at the half dozen or so projects that exist. The Norwegian organisation called the Sahara Forest Project has established the two tiny but innovative projects in Qatar and Jordan. The Sahara Forest project, as its name implies, has very much greater ambitions, as did Charlie Paton the original inventor of the seawater greenhouse. What if we invested trillions of dollars into all the soil and water conservation techniques we have been discussing, and added solar desalination, seawater greenhouses and agrivoltaics to the mix?

*

Farming Deserts

Most deserts have some oasis where life flourishes due to the local availability of water. The abrupt contrast between fertile oasis and arid desert is often sudden: I have stood with one foot on the green and fertile Nile valley and the other on the arid sands of the Sahara. Given water, crops can be grown, people live and animals pastured. Fertile soils can be built up, providing new opportunities for life, and sequestering carbon at the same time.

Desert dwelling communities have long known about and built some truly amazing artificial oases using underground irrigation channels, such as Qanats, to bring water from great distances.[103] The Agdal Gardens in Marrakesh, Morocco, is a walled orchard and garden extending to 340 hectares that is nearly one thousand years old. The irrigation water is brought from the High Atlas down to the gardens along a network of underground channels. Within the garden grow a wonderful mix of trees including dates, olives, almonds, figs, pomegranates, lemons and oranges. We could design and build many more such walled orchards across the region using modern solar desalination, pipes and pumps.

I have previously enthused about the Sekem project and how it has been so successful in turning desert to productive farmland, and achieving a broad range of social and ecological improvements, and how it could be a template for some of our network of model farms. Sekem has used groundwater to kick start their process of desert reclamation. In many areas water tables are falling due to over abstraction. This is why as we envisage the massive scale replication of projects such as Sekem, or the examples of Wadi El Ku, Machakos and the Loess Plateau, solar powered desalination will become a crucial part of the mix, in many of the driest locations.

Solar Power in the Middle East

Earlier in this book I have discussed how various solar technologies are becoming the cheapest form of electricity available to humanity. I have also discussed how, as with the demographic changes of the nineteenth century when people moved to coalfields, in the twenty-first century people will move into deserts. With my proposed Global Green New Deal the idea is to maximize a diverse range of benefits and speed up the transition from 'The Fossil Fuel Age' to 'The Solar Age'. If Manchester, the Ruhrgebiet and Pittsburgh are synonymous with that earlier age, where are the places that are emerging as centres of the solar economy? Perhaps the best already existing example is the Moroccan city of Ouarzazate.

Ouarzazate is a city of 71,000 people located just to the south of the Atlas Mountains and is sometimes referred to as the 'gateway to the Sahara'. Just ten kilometres from the town is Ouarzazate solar power station.[104] It has 510 MW of concentrating solar power, making it currently the largest concentrating solar power plant in the world. It has been built in stages, initially utilizing parabolic trough design, then adding a solar tower with its field of heliostats. Electricity can be produced long after the sun has set due to the use of molten salt energy storage. Later 72 MW of photovoltaic panels were added, bringing the total capacity of the plant up to 582MW. Masen is the Morocco agency responsible for developing the country's renewable energy, and they have worked with a range of technology companies to build this whole solar complex. I really like the way they are adding new and innovative technologies, such as that developed by the Swedish company Azelio using molten aluminium and Stirling engines to store solar heat and convert it to electricity whenever it is needed.[105] This 582MW power solar power station provides far more electricity, twenty-four hours a day, than the city of Ouarzazate needs. The surplus can be sent via the grid to the rest of Morocco, or indeed on through the interconnector to Spain and into the European grid.

The idea of utilizing the vast solar resource of the Sahara for both local consumption and for export to Europe has been around for a long time. The Desertec organization was founded in 2009 in order to develop this concept on a global scale. It was an idea backed by many European companies and entrepreneurs and by many governments and agencies in North Africa. Many interesting projects were proposed, but sadly progress has been slower than many of us enthusiasts would have wished.

Renewable energy resources, and especially solar power, are being developed across the region. There are some huge projects, like Egypt's 1,650 MW Benban Solar Park which just uses photovoltaic panels with no on-site energy storage. In Dubai the Mohammed bin Rachid Al Maktoum Solar Park utilizes a similar diverse range of solar technologies to those used at Ouarzazate, but with plans to build 5,000 MW of solar power by 2030.[106] There are of course many smaller scale projects, right down to the isolated off-grid solar panels fixed to the roofs of simple houses and shacks. Interestingly, and perhaps surprisingly, war-torn and poverty stricken Yemen currently has the highest proportion of its electricity coming from solar power of any of the countries of the Middle East. However, this is mainly due to its tiny electrical usage, rather than massive solar generation.

The deployment of solar power will inevitably rapidly increase everywhere. One of the key tasks of my proposed Global Green New Deal is to explore how this process could lead to the most rapid decarbonisation imaginable while also generating a diverse range of other benefits.

Dreaming Up a Project: The Gulf of Aqaba City-Region

Now, let our imaginations run wild. Let us think how to put together our pieces of that jigsaw puzzle and reveal a picture of a better future. The populations of the hot and arid regions of the world will massively increase over the coming few decades due to naturally growing local population plus in-migration as people seek out new solar powered economic opportunities and tragically flee the inevitable climate related dangers such as rising sea levels. Let us paint a picture of some of the new settlements that might emerge and be attractive destinations for people to flock to. Some new cities will emerge. Some might be located in areas where currently very few people live, and others will be built as already well established settlements grow. Huge projects designed to promote peace and co-operation might profoundly change some places in very positive ways.

The city of Strasbourg is now in France, but its suburb of Kehl is in Germany. Over the last couple of centuries Strasbourg has changed hands several times as France and Germany have fought over territory. Now it is a symbol of successful co-operation, of peace and prosperity. At the northern end of the Gulf of Aqaba sits the Jordanian city of Aqaba and the Israeli city of Eilat. Eilat currently has a population of 52,299 and Aqaba 148,398. Can we imagine this growing to become a huge city region spilling over into north-western Saudi Arabia and also into Egypt's eastern Sinai, spanning these four countries and uniting them in shared peace and prosperity? If one stands at the head of the Gulf of Aqaba it is possible to see the coastline of all four countries at once. They naturally fit into a single geography, and once local Bedouin people would have moved freely across the region.

The envisaged new city might have a population of over ten million, where now less than a quarter of a million people live. So instead of thinking of these as four countries locked in conflict let us apply our model of co-operating decentralized communities. We could think of the existing towns of Eilat and Aqaba being one single community and, conveniently, together their populations total about 200,000, our nominal average sized unit. Let us think that they are enthusiastic to take this concept forward, and to be the locus of in-migration, attracting additional funding commensurate with the increase in population. So if this area grew to be a city region of ten million that would be fifty of our local community units, and if each one was eligible for funding of a billion, or even just one tenth of a billion, together they would attract annual funding of between 5 and 50 billion dollars for this new transnational city-region.

I have previously discussed how the Global Trust for People and Planet might create many millions of new jobs and the training and education suitable for creating a workforce capable of doing what the world needs. If we imagine our city of Aqaba-Eilat becoming an epicentre for this education and training, the first point of investment ought to be in education. Building schools and colleges where Jordanians, Israelis, Palestinians and refugees from war torn countries such as Syria and

Yemen, and the sons and daughters of European and other economic migrants could all come together and build good community relations, would be a powerful first step. A new university would be crucial, and it could have several key departments.

A school of solar engineering would work with partners to build a solar power station like that at Ouarzazate, but much larger and with a much greater emphasis on research, development and deployment of some of the cutting edge solar technologies I discussed earlier in this book. It would seek to design and build a network of solar power stations and projects so that rapidly the area transitions to become a zero emissions region, halting the use of fossil fuels and generating sufficient solar generated electricity and hydrogen to run its growing local economy while also allowing for considerable energy exports. Hundreds of thousands of people, from apprentice solar installers to post-doctorate researchers would be required to take this forward, and the envisaged university would act as a hub for their training and for project design and development. All around the world at the moment there are many tiny start-up businesses seeking to develop innovative designs and uses of solar power. This envisaged university would seek to draw them together and help them all develop and learn from each other. Solar power will be a massive future industry and it will need millions of well trained people to make it happen with sufficient speed to help replace fossil fuel use and avert climate catastrophe. Most of this new development of solar power would be well integrated with improved systems of food production, as we have discussed in the section on agrivoltaics, so helping improve food security, soil formation and creating a virtuous circle of improvements.

The already existing tiny seawater and desert farming project run by the Sahara Forest organisation just outside Aqaba would be massively expanded. This expansion could encompass thousands of hectares of farmland, modelled in part by what the Sekem project has done in Egypt and on the Agdal walled orchard gardens of Marrakesh while learning from previous seawater and saltwater greenhouse projects and Charlie Paton's current shade netting experiments in Somaliland.

Solar powered desalination can involve many different methods, from solar stills to using solar generated heat via concentrated solar thermal systems, solar generated electricity via photovoltaic systems and seawater greenhouses. There is a huge need for more research, development and deployment. It ought to become a huge new industry, bringing life to the world's deserts. Our model farm in the Aqaba region would be trialling them all.

Many thousands of apprenticeships and jobs could be created on this farm come research and training project. Like the Wageningen University and Research Centre in Holland it might spawn many new start-up businesses pioneering many of the things I discussed in the food, farming and biodiversity chapter. It would generate a good quantity of fresh local organic produce for the growing population, and surplus for export, all off a relatively small area of what is now desert, using solar power and seawater as the two key inputs. Fruit, vegetables, nuts and seeds might be the main crops, but Solein, the farm-free food derived from a hydrogen-oxidising bacterium, might become important. So too might aquaculture in the Gulf of Aqaba, with shellfish and seaweed based foods as key products. Insects and fungi may become key parts of the human diet and this may be the place that pioneers their greater use. Students from the university would span out across the globe studying the most successful projects such as the soil and water conservation projects we have mentioned in Sudan's Wadi El Ku, the Machakos area of Kenya, the Loess Plateau in China, Africa's Great Green Wall and embryonic ideas such as those the Weather Makers have for northern Sinai. The decrease in violence achieved along the Wadi El Ku might be a model for productive peace-building efforts across the wider Middle East and especially within the envisaged Aqaba-Eilat city-region. New ideas may emerge, perhaps North Africa and the Middle East could benefit from a Great Green Wall, or walls, and certainly the region could do with many new oases fed by solar desalinated water.

As the city-region of Aqaba-Eilat develops, alongside the Global Trust for People and Planet, they would both seek to have a powerful beneficial impact across the greater Middle East region. A large and first-rate teaching hospital would train medical personnel for the growing city and for outreach into the wider region's impoverished areas, war zones and refugee camps. Schools of governance, of active peace-building, of accountancy and of education would all have important roles to play in the development of better governance across the region.

The region currently, like so much of the rest of the world, is hamstrung by greedy, short-sighted and often despotic systems of government. Corruption is rife and inequality extreme. Reversing all these things is absolutely fundamental to establishing a better future. By setting in place upper limits to wealth accumulation and higher level earnings, and by providing many millions of secure and well-paid jobs, apprenticeships, training and education places, inequality could rapidly be overcome. In many ways we would be reversing the social hierarchy. Anyone with excessive wealth would automatically become subject to investigation as it would be probable they would be guilty of tax evasion and therefore liable to prison. Refugees and economic migrants now so commonly reviled and rejected by countries and sometimes held in horrible detention centres would be seen as perhaps humanity's greatest asset. They are very often motivated by a desire to provide a better life for their children than they had themselves and are prepared to travel to new locations in order to achieve this, and to work hard and to learn new skills. Surely exactly the people we need most when envisaging our new desert based and solar powered cities.

Replicating Solar Cities

In the late nineteenth century New York was booming as a result of rapid in-migration. The Statue of Liberty was unveiled in 1886 with Emma Lazarus's poem cast in bronze inside it. Her poem is often quoted:

"Keep, ancient lands, your storied pomp!" she cries
With silent lips "Give me your tired, your poor,
Your huddled masses yearning to breathe free,
The wretched refuse of your teeming shore.
Send these, the homeless, the tempest-tost to me,
I lift my lamp beside the golden door!" [107]

As the era of Donald Trump and his ludicrous wall subsides into history let us imagine a welcoming and collaborative project encompassing the border between the USA and Mexico. Soil and water conservation techniques could be developed across the whole area, and especially on the entire catchment of the Rio Grande. The American southwest and northern Mexico is an ideal location for the development of solar power. Solar desalination could be used to clean up polluted water, or to desalinate sea water from both the Gulf of Mexico and the Gulf of California. The city of Nogales is half in Mexico and half in the USA. Currently the Mexican part is very much poorer and the border tightly controlled to minimize so-called illegal immigration into the USA. Let us imagine that the two halves of the city wanted to unite and to grow together to develop an inclusive city with free movement of people and ample jobs, education and training for all. So much of what we have discussed in relation to the Aqaba-Eilat project could be replicable in a greater and unified Nogales, or indeed in many other locations along the USA-Mexican border, and elsewhere.

South-western Africa has a very good solar resource which could be jointly developed by Namibia, South Africa and Botswana. Namibia currently has a population of about two and a half million. It would make an excellent location for a solar powered megacity, perhaps where the Namib Desert meets the Atlantic Ocean making an ideal location for solar powered desalination. A few weeks after writing this paragraph the news came in that Namibia is planning to be a major exporter of green hydrogen, in the form of green ammonia.[108]

Chile's Atacama Desert receives the highest solar irradiance of anywhere on Earth, being a relatively high altitude and dust-free hot dry desert. Chile is already beginning to develop this vast solar resource, and with the kind of realignment of the global economy envisaged in this book it could become the epicentre for a solar megacity. As its solar resource is so vastly bigger than anything that would ever be required for all energy use within the country, it is a prime place for the export of solar energy in the form of hydrogen, ammonia or methanol. The Chilean port city of Antofagasta currently has a population of about

402,000 but may make an ideal location for one of our envisaged solar megacities.

Australia also has an impressive solar resource and has had many academics and advocates of solar power going right back to the 1970s. Unfortunately it has had a series of governments utterly in the pockets of the fossil fuel industry and did not build the hoped for concentrating solar thermal power stations that many of us had long advocated. It is now, belatedly, building photovoltaic systems on rooftops and as large utility scale projects. Australia has many places that would make prime locations for a new solar megacity. The Upper Spencer Gulf is one such location and as two old coal-fired power stations were due to close in the vicinity of Port Augusta there was a campaign back in 2012 to replace them with concentrated solar power.[109] Also in the same neighbourhood is the largest and only commercial scale seawater greenhouse in the world. An ideal base to build out from with the range of technologies our solar cities would utilize.

An area of the central and eastern parts of the Sahara has a particularly powerful solar resource, spanning the border of six countries: Algeria, Niger, Libya, Chad, Sudan and Egypt. Might they co-operate to develop some large and collaborative project? It would present challenges as the area is a long way from any coast from which seawater could be taken for desalination, but in the longer term it is conceivable that water might be brought from the Mediterranean or Red Sea for this purpose. Alternatively Egypt might use water from the River Nile. In the 1960s the building of the Aswan Dam created Lake Nasser, and an overflow channel called the Sadat Canal has been built so that in times of high water levels the excess water drains off into the desert, and since the 1990s this has resulted in the formation of the Toshka Lakes.[110] Egypt now proposes extending the Sadat Canal to the Kharga Oasis. The huge Benban solar park has recently been built not so very far away. This region might be ideal as a starting point for some larger solar project, perhaps linking in with neighbouring Sudan, Chad and Libya.

The Thar Desert spans the border between India and Pakistan and would make an ideal location for a peace-building collaborative project to develop the great solar resource of the area. Seawater for desalination could be extracted from the Gulf of Kutch. Near Pakistan's border with Iran the new port-city of Gwadar is currently being built with massive investment from the Chinese. If the mountainous deserts of Balochistan are to be developed, Gwadar could become the city at the core of a co-operative project incorporating parts of Afghanistan and Iran. The Balochistan region of Pakistan currently has a separatist movement and is the scene of much unrest. At present there are so many conflicts and tensions within and between the five countries of India, Pakistan, Afghanistan, Iran and Iraq. This whole region could do with a huge collaborative peace-building effort. Might this be another region, as we have envisaged for the Middle East, for a collaborative peace-building project on a par with the post-war formation of the EU? Again, as with the Middle East, this is a region with ample sunshine and handy coasts from which to extract seawater for desalination. As the EU used coal and steel collaboration to build peaceful co-operation in the years following World War Two, might these two overlapping regions, the Middle East and this region of south and southwest Asia, use solar power projects as the initial building blocks of vastly greater peaceful collaboration?

Let us imagine the collaborative co-evolution of a couple of dozen or more solar megacities scattered around the possible locations I have just discussed in the Middle East, the Americas, Australia, Africa and south-western Asia. They could provide homes and opportunities for hundreds of millions of people all engaged in doing the things that would be most useful to create a better future. They would be training people as teachers, doctors, nurses, engineers, accountants and all the many necessary roles that civilized society needs. They would most importantly be using no fossil fuels and instead generating massive surpluses of renewable energy to help older cities and more difficult sectors of the global economy to decarbonise. They would be pioneering the ecological restoration of the hinterland surrounding each of these cities and developing super productive systems of growing the food that they need without the use of chemicals and fossil fuel derived inputs. Agrivoltaic systems and greenhouses would be at the heart of this intensive and sustainable food production. Peaceful co-operation and the free movement of people on a global scale would be essential aspects of their co-evolution. As Emma Lazarus's poem implied, the poor, the huddled masses, the homeless and the wretched refuse of humanity, so reviled by our current system, could become our greatest asset. They are the doctors, engineers, academics, cooks and bus drivers who will be utterly invaluable in creating a better future. They are worthy of massive scale investment.

Neom Critiqued

Existing governments are beginning to see some of the same opportunities as I have been exploring, but their interpretation of what to do and how to do it is in many ways the polar opposite of what I am advocating. They talk about a zero carbon economy without seeming to have much clue about what that means, and they totally ignore the absolute necessity of radically reducing inequalities. Their plans seem to seek to perpetuate and even increase inequalities without realizing that this will only exacerbate injustice, resentments, social division and conflict. They often seem fascinated by high tech frippery, not realizing which technologies are vital and which are nonsense.

Mohammed bin Salman, crown prince of Saudi Arabia, is promoting a huge $500 billion project for a new city region in the deserts of north-western Saudi Arabia to be called Neom.[111] The concept is intriguing in that it purports to be a zero carbon venture, powered by renewables, with desalination and a whole raft of high tech, including super modern transit systems and walkability, with no cars, all designed into the layout. So far so good, but it seems fraught with contradictions. It proudly announces the flying times to many great cities of the world without mentioning that zero carbon flights to such destinations do not now exist, and may not for a very long time, so rather undermining the claims it makes to be a zero carbon development. It highlights that it will have a great concentration of Michelin-starred restaurants without saying how it will contribute to feeding the vast majority of the local or the world's population who've never eaten in a Michelin-starred restaurant. The Saudi record on human rights is appalling, including their ousting of the local Howeitat Bedouin tribes people from the Neom site, their brutal intervention in the Yemeni civil war and the assassination of Jamal Khashoggi. Many of the global companies that the Saudis are planning to work with may not want the reputational damage of involving themselves in this project. The plan seems to be based on the assumption that the form of capitalism shaped by market fundamentalism and ever increasing inequality will go on forever.

The project proposed by Mohammad bin Salman includes some technologies that have not yet been invented and some that seem pointless frippery, such as flying cars, robot dinosaurs, robot maids and an artificial moon. The project also includes some of the same technologies as I advocate, and which will be invaluable in transforming the Saudi economy. Until very recently Saudi Arabia generated almost all its electricity by burning oil, and its economy remains vastly dependent on oil exports, which given the low price of oil over recent years has meant that the Saudis are often pumping oil at a loss. For the sake of their own economy and for the global environment they need to rapidly develop their huge and almost totally untapped solar potential. They have just recently inaugurated their first large scale solar power station, the 300MW solar photovoltaic power plant at Sakaka in the north of Saudi Arabia, just outside the area planned for the Neom development. There are plans to build a lot more renewable energy infrastructure including the world's largest green hydrogen production facility.[112] Time will tell which, if any, of Mohammad bin Salman's ideas actually get developed.

One of the reasons Neom is attractive is that it is thinking how major investment can be made to reduce the Saudi economy's dependency on the oil industries. Mohammad bin Salman has proposed an initial $500 billion investment. I think, given that budget, we could do very much better. If the kind of Global Green New Deal that I am advocating was enacted, despotic leaders like Mohammad bin Salman would see their power and wealth vastly decrease, and the class, gender and power structures of Saudi society change beyond recognition. We could forget about the high tech frippery and concentrate on what really needs to be done: decreasing Saudi oil dependency and carbon emissions, improving local food production, providing better training and education, and much improved human rights and social freedoms. Given the diminishing role of nation states and the enhanced role of networks of local communities working together to solve common problems, perhaps we can look at this region of north-western Saudi Arabia as an extension of our project focused on Aqaba-Eilat, but spilling over into Egypt and Saudi Arabia.

Emergent System Change

Dear reader, you may at this stage be wondering how on earth do we get from where we are now to this imagined future with its Global Green New Deal and massively ambitious plans for global social and climate justice, radical political decentralization and global solidarity. No government, United Nations agency, think tank or any political party is as yet proposing anything like this. Where will leadership come from?

System change will come. It may, or may not, involve some kind of Global Green New Deal. It may, or may not, involve something like my envisaged Global Trust for People and Planet. We don't know. Nobody does. System change could mean going from what we have to some kind of hyper-capitalist global totalitarian state with a tiny all-powerful elite and a system of global slavery with whole populations bought and sold. There are any number of dystopian possibilities. Tragically there are many politicians and powerful people whose pursuit of ever greater personal wealth and whose ideological belief in their own superiority is pushing us towards such a horrible future. But things could work out well. Positive change could happen. If we make it happen. Major changes, be they social, economic, technological or political, usually are the result a lot of effort.

I have said that the system change so many of us desire will not be a top-down process, not initially. It is already starting in myriad ways all over the world, often drawing on deep roots in long established ideas and systems. Some things that are as yet tiny may soon mushroom into very much bigger and more influential ideas and projects. I have mentioned dozens of technologies, land use systems and projects that I see as being a good basis for building something better. There are very many more that I have not mentioned. Together they are already, or have the potential to be, the seeds of system change: very positive and beneficial system change.

By writing about major change toward a better, more socially just and ecologically healthy future, I hope to be in some small way contributing to its creation. Writing about these concepts does help spread ideas, and these might trigger someone, somewhere, with vastly greater resources than I have to act in a more impactful way than I could possibly manage. How might something like my envisaged Global Trust for People and Planet emerge?

Let us just start by looking at money. There is a lot of money sloshing around. Many people have bank and building society accounts getting poor rates of interest. Sovereign wealth funds and ethical investment funds are always looking for good projects to fund. Governments have wasted truly vast sums on quantitative easing and so very much else. What if the people responsible for all this money saw a project that was financially quite safe, offered a reasonable rate of return and was clearly highly ethical, doing important work to help people and planet in many socially and ecologically beneficial ways? Some, but not all, would surely invest in such a project.

Social and ecological activists have often been hamstrung by thinking too small. They often worked on the basis that money was hard to find and that projects needed to be small scale and rooted in local communities. There is huge competition for grant funding measured in a few thousand to a few million pounds. If we think of a project measured in billions, or hundreds of billions of pounds, there will be very little competition for funding. There will not be grant giving bodies set up to award such sums, but there are many other potential sources of funding.

At this stage in human history we need to be very much more

ambitious, and many factors seem to me to be on our side. There is a crying need for impactful action on the climate and ecological crisis, and also on the social and economic crisis. There is strong demand for energy, and solar is now the best way to meet this demand in most geographical locations. There is potentially an enormous amount of money looking for an investment opportunity that is safe, ethical and pays some kind of economic return. Could we set up some kind of large scale solar project that was deeply committed to improving the lives of some of the poorest people on the planet, and also to achieving a broad range of other social and ecological goals, all while being profitable?

In the section on rewilding we looked at how Jeremy Leggett is setting up a mass ownership company called Highlands Rewilding Ltd as the vehicle to buy land across northern Scotland with the intention of creating new forms of employment based on creating a richer ecological habitat than currently exits. He is not himself from the Highlands of Scotland, but is setting up the project to benefit the people of the region, and after the initial ten years, or in the event of him dying, the ownership of the project would go to a Trust. He envisages several rounds of fundraising. He has raised a few million pounds and intends to raise a few million more. What if it was massively oversubscribed? Could the principles be taken to land elsewhere?

Building Societies grew from the late eighteenth century until the late twentieth across Britain. These mutual societies have many advantages over banks and have suffered less from mismanagement and corporate greed. Triodos is one of the few ethical banks, and it has been better managed than any other bank that I can think of. Now might be the time to institute some kind of structure that draws on the tradition of mutual building societies, ethical banking and mass ownership of land as pioneered by Jeremy Leggett and his team. This new institution might want to work with a number of industrial partners to develop an initial project that is both profitable and ethical. If that one is successful they might then want to develop many more.

There are many millions of people desperate to see big scale change that offers hope that we can turn things around, simultaneously taking action on many aspects of the climate/biodiversity/social crisis we all face. Countless charities, development agencies and individual citizens might want to put time, energy and money into such a project.

As I have argued throughout this book solar power will become the biggest energy source as we move away from fossil fuels, and the hot dry deserts and semi deserts of the world are where this can be done most cheaply. This solar electricity will be stored in the form of green hydrogen, methanol and ammonia, and these will become globally traded products. This is all beginning to happen and governments and industrial businesses are enthusiastic to get projects off the ground.

I can think of hundreds of potential projects, but I do not have the capacity to take any of them forward. It would take a team of people; some with experience of raising funds, others with experience of initiating projects, and some might be charities, some industrial companies. Whoever makes up that team will shape the initial project.

I have speculated about the possibilities of massive expansion of renewable energy co-operatives. In the section on co-operative models I enthused about Awel Aman Tawe which set up the Egni solar co-op, the Awel windfarm, and does lots of great community work across Wales. I also enthused about larger scale municipal ownership and specifically about the 400MW Trianel Windpark Borkum. I enthused about the possibilities of a global renewable energy co-op, with all 7.8 billion of us as members. The full-on global co-op is a long way off, and if it is ever to come about it would be in stages. What might the next steps be to get from where we are to where we'd like to be? Like most things, it might begin with a conversation. Who would be the best people to have in the room, or on the Zoom call, to initiate something along the lines of an international solar co-op, initially building one, two or half a dozen projects, but with the desire to go very much bigger?

An Imagined Zoom Call

Let me sketch out an imagined Zoom call I would love to play a part in. On the one hand it would be great if some key people from the world of renewable energy co-ops were involved, such as those running our local co-ops, Jon Halle from Sharenergy, and Dan McCallum from Awel Aman Tawe. We would need someone from RESccop, the organization that co-ordinates and promotes renewable energy co-operatives across Europe, and someone from Solar Power Europe, the organization that does what its name suggests, and promotes the greater use of solar power in Europe; and no doubt some of the global renewables advocacy organizations would have much to contribute.

Jeremy Leggett's recent experience of initiating mass ownership of land with social and ecological goals would be most useful, as would his previous experience of setting up Solar Century. Trianel, the organization that supports European municipal energy and infrastructure projects, would have much to contribute. I have said that combining solar power with good farming systems will be important, and so having someone like Byron Kominek from the Colorado Agrivoltaic Learning Centre would be useful.

It would also be useful to have some big companies with strong ambition to develop renewables and also to work in ways that are socially beneficial. Statkraft is currently Europe's largest renewable energy producer and is rapidly expanding into global markets. Siemens is another global player with a strong interest in greener technologies, and is at the forefront of some green hydrogen projects. Having someone from banking would be useful, with Triodos the most obvious choice.

A logical first step would be for some of the larger and better funded organisations to provide some initial funding to employ a few people to work on developing a small organization to take things forward. An initial project might be anywhere in the world. The goal would be to develop larger renewable energy co-ops along lines that were replicable on a global scale. Projects that had the potential to create a broad range of benefits would be selected. Each project would seek to be profitable, so that investors would get some return on their money, and a generous pot of money could be used to help improve local food, water, energy and economic security for the local population wherever a project was located. Our imagined global co-op might want to look at a number of potential projects and partners, and select the most easily achievable project to start with. Let me float half a dozen or so potential projects. Each would already probably be potentially viable, even without my envisaged Global Green New Deal.

Half a Dozen Potential Projects

1.   Statkraft is already active in Chile and Peru. In Peru it is developing hydro electric power and is working with local communities to irrigate orchard crops with modern drip irrigation in the dry lands of western Peru. Statkraft stresses its socially responsible basis for this work.[113] This looks like the ideal sort of place to start to develop agrivoltaic systems to meet local power supply, to save precious water and increase fruit and vegetable production. Statkraft already has the staff working in these local communities. It would be a good place to start in a small way with a few medium sized agrivoltaic systems with a strong element of education and research, perhaps modelled on Jack's Solar Garden and the Colorado Agrivoltaic Learning Centre.

The semi-desert coastal strip of southern Peru, and the true desert of Chile's Atacama would be an ideal place to do larger scale solar power. The left leaning Gabriel Boric has recently become president of Chile, and he might be supportive of a co-operative structure. Somewhere in this region might be a very good location to set up a very much larger, probably multi-GigaWatt, project. It might include concentrating solar power with thermal energy storage, and solar photovoltaics and battery energy storage. It could be a centre for the export of green hydrogen. The Atacama gets the highest solar irradiance of any location on Earth, so solar power has the potential to be very profitable. There are many mineral mines in the area which need reliable electricity supplies and are sometimes a long way from existing grid connections. They would make ideal initial customers. Some kind of citizens assembly representing local communities might be useful to assess how best to invest money to help create new and useful projects. Many new training opportunities could be created for solar engineers and farmers to work together to optimize how best to use agrivoltaic systems in their localities. Statkraft are already active in Chile with wind and hydro projects. It seems highly likely that they will develop large scale solar power soon anyway, so I would suggest to them that they might like to consider some kind of co-operative model endeavouring to develop their existing social responsibility into something very much more ambitious. I think many investors and funding bodies would love to put money into something along these lines.

2. Morocco has been at the forefront of the solar power revolution, and the achievements of the MASEN agency and the especially the establishment of the Ouarzazate solar power are to be celebrated. Morocco is rapidly bringing electricity to the vast majority of its population, so rapidly in fact that although the use of renewable energy has increased so too has the use of fossil fuels, including coal.[114] More needs to be done to ramp up renewables. The region encompassing Morocco, Mauritania and the Western Sahara, rather like the Middle East, is beset with political problems and extraordinary potential. Morocco and Mauritania have both made territorial claims to the Western Sahara and many of the population of Western Sahara live in tented refugee camps in Algeria. There are parallels with the situation of the Palestinians in the Middle East. A vast minefield runs across the Western Sahara.

The potential of the region is enormous. It has excellent solar and wind resource. The sea routes to Europe are relatively short and the potential for increased sustainable trade and co-operation very exciting. One of the largest iron ore mines in the world is located in the deserts of northern Mauritania, and linked to the port of Nouadhibou by a railway. Currently the iron ore trucks return empty from port to mine. They could be used to transport all the equipment needed to build a large solar power station which initially would be used to provide the mine with electricity, and to electrify the railway, and as it grows solar powered iron ore smelting and steel making could be developed. Sweden is currently at the forefront of green steel production using renewable energy and hydrogen. A partnership between Sweden and Mauritania could do much to bring such a project forward. Widening the partnership to include all the five Nordic countries, linking them with Morocco, Mauritania and the Western Sahara could lead to very significant development of solar power in this hot arid region, upon which peace, prosperity and co-operation could be built. Robert Habeck, appointed in 2021 as minister for climate and economy in Germany, is committed to lowering German carbon emissions. He has ambitious plans for renewables within Germany, but it seems to me Germany might usefully import renewable energy from both the Nordic region, and from this region of Morocco, Mauritania and the Western Sahara.

Germany and the Nordic region are at the forefront of co-operative models of renewable energy generation. Might they work with Morocco, Mauritania and the Western Sahara to develop a number of large solar projects in this part of the Sahara, with a view to helping solve multiple problems including Germany's high carbon emissions and Saharan poverty and conflict? Very large scale solar ought to be highly profitable in this region, so making funds available to help local people and projects, and give a reasonable return to investors and industrial partners.

3. India is a bit like Morocco in that it has invested in renewables, but coal use has also grown as the increased use of electricity has been rapid. Like Morocco it has a great solar resource. As I've discussed earlier the Thar Desert on the borders of Pakistan and India is the obvious place to develop a very large solar project. Could it be along co-operative lines, and profitable already, without any of the mechanisms for tilting the field toward a low carbon future as envisaged in my Global Green New Deal? I think there is great potential.

Uttar Pradesh is India's most populous state. Along the wide and fertile Ganges Valley coal fired power stations have been built and more are planned. This would be an ideal region to trial large scale agrivoltaic systems. As has been shown by the tiny Colorado Agrivoltaic Learning Centre, reducing the need for water to irrigate crops under the solar panels would be tremendously beneficial, and the solar panels would already produce electricity more cheaply than the new but already obsolete coal fired power stations.

4. Most of the billion or so people living without access to electricity live in rural Africa. Decentralized solar, with batteries and micro-grids, is how most of these people will get access to energy. It is beginning to happen very quickly. A couple of new start-up companies are at the heart of developments. Energicity was set up by two Americans, Nicole Poindexter with a background in finance and Joe Philip an electrical engineer with a background in solar. They are initiating projects in Ghana, Sierra Leone, Benin and Nigeria. Bboxx is a similar start-up operating in eleven countries in Africa and Asia, which was set up by three students from Imperial College in London. Both companies are utilizing a number of digital technologies, solar panels and batteries to bring many benefits to the lives of poor people in innovative ways. They are well integrated into the markets of rural Africa with local staff. They might be ideal partners to develop a whole network of thousands of agrivoltaic systems modelled on Jack's Solar Garden and the Colorado Agrivoltaic Learning Centre. In the dry savannahs of Africa the water saving characteristics of agrivoltaics could be extremely useful. Might campaigning and charitable organisations, from Avaaz to Oxfam, want to join such a project? I would love to see pioneering African farmers such as Patricia Kombo and tree nursery founding climate activists such as Elizabeth Wathuti involved. Combining the best mix of solar, digital and battery energy systems with the best ecologically regenerative farming systems has huge potential. This might not be as directly profitable a project as say a vast solar and green hydrogen project in the Atacama, but it might still be profitable, and maybe profit does not need to be made on every project. In terms of helping a diverse range of problems this might be one of the best projects imaginable.

5. The situation in China is critically important, given its continuing vast use of coal. The Chinese have massively ramped up renewables already. Their energy demands have risen rapidly over recent decades as they have become the global centre of manufacturing. The global change from a linear throwaway economy to a modest and circular economy would see demand for Chinese made goods contract. China will need to create more jobs if this happens, as happen it must. China could establish large solar power projects in the Gobi Desert, as it has already started to do, but this could be massively expanded, and include some solar projects high on the Tibetan Plateau, which has a remarkably good solar resource. Across the bulk of China agrivoltaics has great potential.

    Chinese manufacturing is well geared up to run this without any external help, and it has the necessary financial resources. However, international collaboration would in itself be useful. In terms of thinking about an international co-operative model for renewable energy I think it unlikely that China would be one of the first to sign up, but if they did it could massively amplify the impact of any such project just because of the huge scale of China's financial and manufacturing sectors. It is certainly in China's interest that humanity rapidly reduces its carbon emissions. China is very exposed to risks due to glaciers melting, droughts, floods and sea level rise. International co-operation on reducing emissions, on developing renewables and on much else would greatly benefit from active Chinese backing. Their Belt and Road network of massive global development projects could be tilted more towards developing renewables.

6. The previous five suggested possibilities have all focused on a few key markets for solar power. Of course there are other locations, notably Australia, which might be at the forefront of a global energy co-operative based on solar power. I want, with this sixth possible area, to look at the northern and western fringes of Europe. The Nordic Region plus Scotland and Ireland and the seas all the way from the Atlantic to the Baltic are going to host a massive expansion

of offshore wind farms. This is already happening. The 40MW Middelgrunden and the 400MW Trianel windfarm are two of the largest co-operatively owned renewable energy projects in the world. Now is exactly the time to further ramp up ambition and think of another ten-fold increase. A 4,000MW, or 4GW, project in the North Sea linked into the grids of Denmark, Germany, Norway and the UK would instantly be profitable, and many organisations are already pushing for this kind of development, but none as far as I am aware is thinking of doing so under a co-operative model of ownership. Now is the time for organisations like Trianel to be thinking bigger and looking for partners to develop a larger scale co-operative project.

If a sufficiently good structure was created to do an initial 4GW co-operatively owned project in the North Sea they might then want to do half a dozen more. Floating wind off Ireland, Chilean and Moroccan solar, an Icelandic interconnector, and perhaps some of the more ambitious peace building co-operative projects I have speculated about in areas such as Mauritania and Western Sahara, or in the Aqaba-Eilat region of the Middle East, might all follow. All this might happen in the world as it is now, but with the global level changes to taxation and the rest as envisaged in my Global Green New Deal it would greatly be helped.

Colder climates

Throughout this book I have been stressing the possibilities of people moving to where energy is cheapest and new opportunities greatest. I have mainly focused on the hot sunny desert regions of the world and their tremendous potential to develop solar power. There are of course many similar opportunities in colder climates where hydro, wind and geothermal resources are greatest. There are a number of such regions around the world. Iceland is one such, with its huge hydro, geothermal and wind resource, and also potential for offshore wind and marine renewables such as wave and tidal power. Tierra del Fuego on the southern tip of South America is one of the windiest places on Earth. I will focus my attention on just two of these colder regions, the northern fringes of the British Isles and the far north of Sweden.

The seas and lands that constitute the north-western fringes of the British Isles are a region with enormous wind power potential. They also have good potential for marine renewables such as tidal and wave power. I shall look at this area later in this chapter, while also looking at my local patch, Herefordshire, and how it fits into the changing situation within the British Isles.

The main cold-climate area I want to focus on is the Nordic region generally and more specifically on the small city of Luleå and Norrbotten County in Northern Sweden.

Luleå, Norrbotten County and Northern Sweden

Luleå is a small city, but still the largest in sparsely populated Norrbotten County in Northern Sweden. Several new technological developments are already happening in this region. It could be argued that it is already an epicentre of 'The Solar Age', despite paradoxically being a very long way from the hot sunny deserts. The Arctic Circle runs across the Norrbotten County. This region is already showing rapid economic growth based on low carbon technologies. These developments have the potential to be highly disruptive for established global high carbon emitting industries based upon the use of fossil fuels. This area is already showing signs of considerable in-migration with huge new opportunities for industrial and urban growth based on cheap low carbon renewables.

Northern Sweden has long had abundant hydro-electric power. Over recent years it has built a lot of wind power, including the still expanding Markbygden wind farm in Norrbotten County. It may grow into one of the world's largest windfarms, possibly up to 4,000 MW. Solar power, perhaps surprisingly, is also emerging as a possible major energy source. Solar panels work more efficiently in cooler conditions, and the long daylight hours and clear skies and not too hot summers in Northern Sweden are ideal for a good, but seasonal, energy supply.

Northern Sweden has some of the lowest carbon electricity in the world. The electricity supply is also very reliable and cheap. A number of companies that want the kudos of using only renewable energy, and the cost advantages of doing so, are moving to the area. The benefits will be especially large for companies involved in processes that use a lot of energy like steel making, battery manufacturing, data centres and making green fuels such as eMethanol.

Currently, globally, steel making is one of the biggest sources of carbon emissions. It is of crucial importance to rapidly decarbonise the process. H2greensteel is a new company, founded in 2020, that is seeking to disrupt the global steel industry. It is building a factory at Boden, near Luleå in Norrbotten County which it is planning to have in low carbon steel production by 2024, and producing five million tons per annum by 2030, supplying so called green steel to many of Europe's big car manufacturers.[115] Norrbotten County has long been a major centre for mining iron ore and gold. H2greensteel plan to use the abundant and cheap renewable electricity to make green hydrogen. Green electricity and green hydrogen will then be used to make green steel, so making the use of coal in steel production obsolete.

Northvolt are another Swedish cleantech start-up. They are currently building a vast gigafactory for battery production at Skellefteå, just outside Norrbotten County, in neighbouring Västerbotten County. Again, cheap low carbon electricity is a powerful draw to operate in this far northern climate. While H2greensteel will be providing the steel to European car manufactures, Northvolt will be supplying the batteries, and of course steel and batteries have many more uses than just in making cars. Currently Northvolt's factory in Skellefteå employs over five hundred people from fifty-six nationalities, and looks set to expand over the next couple of years. This is early evidence of the kind of in-migration to areas of cheap renewable energy that I have frequently discussed.

In neighbouring Västernorrland County, in the town of Örnsköldsvik, the company Liquid Wind are planning their FlagshipONE factory. This will use wind power to make green hydrogen, which will be combined with the waste carbon dioxide from a very efficient combined heat and power plant, and so be used to make green eMethanol. They hope to be producing 50,000 tonnes of eMethanol per year by 2024, which would make it by far the largest such plant on the planet. I would expect many other plants to follow, wherever cheap renewable energy is abundant, and ideally also where carbon dioxide emissions can be captured. The fuel will be used to help the global shipping industry swap from diesel to methanol. The same types of fuel may be used in future to help make air transport greener. Initial supplies of eMethanol will be used to fuel the fleet of new container ships that Danish shipping giant Maersk has just ordered. [116]

Data centres use a lot of electricity, and Facebook built their first data centre outside the USA in Luleå, largely because the low cost, low carbon electricity supply. Many other digital companies are located in the region. Luleå University of Technology is closely tied in with many of these developments and is at the forefront of research in mining, renewable energy, intelligent industrial processes and other related topics. A new high speed rail route is being developed linking Luleå to the towns and cities of the rest of Sweden. The investments in green steel, battery manufacture and data processing all generate money and a basis for further investments, in the university, the new railway and no doubt schools, hospitals and the like. This is the kind of virtuous cycle of investment that might benefit so many poor places in the sunny tropics; and how solar power is developed in these regions will to a large extent map out their path out of poverty, and the world's path out of the climate crisis.

Sweden generates a lot of low carbon electricity. The Swedes also use a lot of energy. They tend to have large houses and heat them to pretty warm levels. They are already leading the world in terms of good insulation, and several pioneering projects are showing how they could become very much more energy efficient. Many houses are connected to very efficient district heating systems, and with their low carbon electrical grid and heat pumps this is another way to go.

One totally solar powered experimental house, the Zero Sun House, in Skellefteå is not connected to the electrical grid or to a district heating network. [117] It uses roof mounted solar panels to generate electricity and store the summer surplus in the form of hydrogen for winter energy use, when it goes for months with little or no sun. The logic is if it can be done here in northern Sweden it can be done anywhere on Earth. This kind of architectural design and construction could be very useful in many remote locations around the world. It is probable that vastly more of the global housing stock will be built with sufficient solar panels on their roofs to be net electricity exporters rather than importers, or, if not connected to the grid, then to be energy self-sufficient. Probably very few will need to use such an expensive option as interseasonal energy storage, but the logic of this house in Skellefteå is if it can be done here it can be done anywhere.

We see here a whole cluster of innovative technologies emerging in remote northern Sweden. All of them have the possibility to be disruptive to the established global fossil fuel industries. Be it year round solar powered off grid houses, green steel, green hydrogen or eMethanol production or a battery gigafactory, all have great potential to create change. System change is emerging in many places and in many ways, and northern Sweden is showing itself to be one very interesting such area. Technologically it is at the forefront of many interesting developments. I can imagine it, under something like my envisaged Global Green New Deal, becoming a great hub of innovation, co-operation and constructive and positive in-migration, even more than it is already.

One critical test of how this new industrial revolution unfolds in northern Sweden will be how it can be reconciled with the traditional Sami reindeer herders. It ought to be possible to balance these competing interests, and Sweden is as well placed as any country I can think of to do this in a fair and democratic manner.

Herefordshire and the Disunited Kingdom

Let us imagine what Herefordshire might look like in the early 2030s. Many of the county's biggest employers will have gone bankrupt or transformed their business model beyond recognition. Once we have stopped using all fossil fuels, what will our transport systems look like? Where will our electricity come from? How will the landscape look and how will farming be different?

The next decade will be a time of momentous change, whether we like it or not. The climate will become more unstable, more highly energised and weather forecasting will become more difficult and less easy to predict. Extreme weather events will be very much more common. We will need to have cut global carbon emissions to zero, and to be drawing down past accumulations of atmospheric carbon back down into the ground where it belongs, and where it can do good.

Brexit will have brought a decade of chaos, increased poverty for ninety-nine per cent of the population and repulsive levels of excess for a tiny minority. The UK will probably have split up, with Scotland leaving the UK and re-joining the EU, and be the first part of the UK to adopt more Nordic policies, starting on the long road to more sensible government, equitable distribution of wealth and ecological recovery. Demand for Welsh independence and re-entry to the EU will be growing, as they watch the progress of the Scots with growing envy. Northern Ireland may well have united with the Republic of Ireland and be back in the EU. England will be deeply divided, poorer, and have lost any role in global leadership. Politically the divisions might be something like Jacob Rees-Mogg leading a dwindling rump of far-right market fundamentalist extremists, and Caroline Lucas leading the majority of the country towards playing an enthusiastic role in the global collaborative process to implement some kind of global green new deal, which may well look something like the one that I am sketching out in this book. Eventually proportional representation will have been introduced, allowing the Labour and Tory parties to split into several diverse parties, much to the relief of most of the members of both. There would be at least a couple of dozen parties in the English parliament, with usually four or five forming a governing coalition. Most would be in favour of following Scotland back into the EU, if and when the EU would accept us back after the damage, chaos and cost that Brexit has inflicted.

Global politics will look very different. Superpowers will be a thing of the past. Networks of co-operation are already and will increasingly become the dominant model. The EU and the African Union will probably be co-operating very much more closely. Climate leadership is already coming from a network of small island states from around the world, with the Nordic region, New Zealand and many smaller countries from Bhutan to Uruguay, Costa Rica to Singapore, each demonstrating leadership in various ways, and learning from each other in a powerfully collaborative way. In 2019 the Wellbeing Economy Governments partnership (WEGo) was formed linking Scotland, Wales, Iceland, New Zealand and since 2020 also Finland, into a collaborative network to promote human and ecological wellbeing. By 2030 might WEGo grow into a powerful force for good in the world, and be of more long term significance than the disaster that is Brexit?

Globally military budgets would have shrunk dramatically, and the remaining units would increasingly be seconded to the UN for global peace-keeping and peace-making, disaster relief operations and for combating terrorism, organised crime and people trafficking. They may even have a role in helping track down rogue billionaires, wanted by the International Criminal Court for tax evasion, or for seizing the assets of companies found guilty of ecocide. For Herefordshire this might mean the disappearance of the SAS, or it may continue as a branch of this UN-led organisation.

By the mid 2030s the importance of multi-national corporations and national governments will have declined. Our MP's importance will be somewhat diminished. Herefordshire County Council will have a very much larger budget and greater responsibilities. It will be engaged in networks of co-operation, something like a vastly expanded version of the Aalborg Process, where best practice on all manner of things could be studied, built upon and standards continually improved. The economy will be very much more localized. There will of course still be global trade, but the proportion of goods and services owned and traded within the county will be very much greater.

The proposed Global Green New Deal would have a huge impact

on pretty much everything. The changes to taxation would effectively eliminate the extremes of wealth and poverty, both globally and also here in our county. Health and education would be very much better funded, and private schools and hospitals would all have closed down, unable to compete with the excellence of the free-at-point-of-use general provision. We would systematically have applied the Nordic model, or more specifically the Finnish model.[118] Children would be happier and more at ease with themselves and with the educational system. They would attend school for fewer hours yet achieve more, due to their increased levels of happiness and self-confidence, based on greater autonomy and mastery of their chosen interests, as we explored in the section on work. Wellbeing, human dignity and equality would be at the heart of everything.

The term United Kingdom might have slipped out of use. It is a region currently defined by its lack of unity. It is certainly a Disunited Kingdom at present. It seems to me that republicanism is growing and it may cease to be a Kingdom. The term Great Britain seems incredibly dated, harking back to a long gone imperial past. As Scottish Independence looks increasingly inevitable one is left with the question about what to call the remaining parts of our island nation. Assuming Northern Ireland also leaves then maybe Wangland as a portmanteau of Wales and England would be appropriate? Or simply England and Wales, which might be better as it implies their probable eventual splitting?

Herefordshire: Transport and the Economy

When life beyond fossil fuels was mentioned a few years back many people used to think either of a return to the horse and cart or to sci-fi type jetpacks. Nowadays they envisage a straight swap from petrol to electric cars. There are a number of important trends and technologies which are poorly understood and which will have a major impact. The need and desire to spend time travelling may well decline very quickly, and it would certainly be a goal of any green new deal. Travelling should become an infrequent but joyous exploration again rather than a boring daily grind.

The Covid pandemic has shown how quickly things can change. Many more people will work from home. Zoom, Skype and other online tools have replaced the need for many businesses to pay the rent on office space, and increasingly physical meeting spaces can be hired by the hour. I see a huge new role for public libraries as free open spaces for people to work, study, record music or to make videos. Much city centre office space may be converted to housing. Many cities will follow Paris in seeking to make as many services as possible available to people within fifteen minutes walk or cycle from where they live. Already car ownership is falling for residents of many cities, and with the right policies it could fall very much faster and further. Walking, cycling and public transport would of course all receive massively increased budgets and road building be stopped. In many urban areas we may already have passed peak tarmac: the trend may be to convert car parks and roads to parks. Utrecht recently converted its urban motorway back into a city moat, as it had originally been.[119] Might Hereford convert Edgar Street and Blue School Street into a beautiful waterway, linking back to the Wye down Mill Street? Car parking spaces in the city could be dramatically reduced. Business rates could be calculated at least in part on the size of a business's car park, which would help reduce the role of supermarkets in favour of small locally owned businesses located in close proximity to their customers.

All children should be able to walk or cycle safely to school. Bike buses, like a pedal powered minibus, are already proving popular in Holland,[120] and may well work successfully in Hereford, once car traffic has been reduced and slowed down. Electric cargo bikes are replacing petrol and diesel vans in many cities, notably in Berlin.[121] In Hereford we have Pedicargo and the Beryl bike scheme leading our transport transition. So much more could be done with the right policies in place.

There are many advantages in sharing resources, and cars are one thing that many people own but only use for a tiny percentage of the time and, with the changes advocated in this book, may use very much less in future. Until recently, Herefordshire had two functioning community owned and run car sharing clubs; Malvern Hills which extends to cover Colwall and Ledbury, and the St James and Bartonsham club in Hereford. I am a member of the Hereford scheme, where about forty or so households share the ownership and use of five cars. It works well. Being a member saves money, cuts pollution, reduces car parking pressure, builds community cohesion and gives greater flexibility in terms of having the most appropriate vehicle for any particular use. With our excellent new Green and Independent led coalition council, help is being provided to establish several more such car share clubs. Of course more could be done. Administrative assistance and training could be provided, so too could assistance to acquire electric cargo bikes and other greener technologies. Our local car share club in Hereford would love to replace all our fossil fuel powered cars, but can only do so as quickly as budgets allow. We already have one electric car, are considering buying a second, and are awaiting trialling a Riversimple hydrogen fuel cell car.

Riversimple is an excellent example of the circular economy. They are, as far as I am aware, the only car manufacturer in the world that plans never to sell a car. When products are made for sale it is in the maker's interests to build in obsolescence, so more products will continually need to be made and sold. That is how the linear economy works. Once the useful life of the car is over it is considered as waste, to be disposed of. With Riversimple, as in many examples of a circular economy, the product is designed to be leased. It is then in the interests of the manufacturer to build in durability and ease of re-use of the car's constituent parts. Even the materials from which the car is made are leased from the suppliers. This pushes the circular model back up the supply chain. This model, applied across the entire global economy could dramatically cut energy and resource use, and the quantities of things going into landfill. Riversimple originated in Herefordshire, and now are located in Llandrindod Wells in Wales.

For cities, towns and larger villages it is easy to imagine a future where most people no longer feel the need to own their own car. In dispersed rural communities public transport is very much more difficult to organize, and distances for walking and cycling are more challenging.

Currently many rural people come into Hereford to do their shopping or to go to work. If more jobs, services and shops were located in the market towns and villages the need to come into Hereford would be reduced. Rural residents might drive to their local market town and then take a bus or train into Hereford if they really needed what only Hereford could provide. Car sharing rather than car ownership can work for some people in rural areas, as Malvern Hills car share club have demonstrated. The number of cars in the county will decrease and an ever increasing ratio will be in shared rather than individual ownership. Battery electric and hydrogen fuel cell will have entirely replaced petrol and diesel.

Railways will have been reinvigorated through new investment and technology. New railway stations may have opened at Pontrilas, Moreton on Lugg and Rotherwas. Mainlines will have been electrified, and quieter lines such as those from Newport through Hereford and up to Manchester, and from Hereford to Birmingham, will have switched from diesel to hydrogen fuel cell. Bus services within the county will have improved and be either battery electric or hydrogen fuel cell. Hereford may have a tram system up and running by the early 2030s, probably something like the new one being trialled in Santa Cruz, California, which uses a hydrogen fuel cell and battery combination, making the installation of overhead cables unnecessary. [122] The number of trucks and vans on the road will have declined as the economy becomes more localized. Heavier long distance trucks and buses will probably be hydrogen fuel cell while lighter and more local vehicles will probably opt for battery electric. For refrigerated transport the Dearman nitrogen engine will provide propulsion and cooling.[123]

The changes to transport policy and provision will have had significant positive effects on public health. Slower speeds, less traffic and better walking and cycling will have reduced traffic accidents, improved air quality and led to more people getting more exercise. As more parks, gardens, cafes and play areas replace car parks there will be more space for convivial interaction. Most people will be fitter, healthier and happier.

Energy: Herefordshire and the British Isles

The British Isles will probably become a net energy exporting region. North Sea wind farms will continue to expand and the seas from the west coast of Ireland to the west and north of Scotland have a tremendous wind resource, where very large scale floating wind will be developed. Interconnector electric cables and hydrogen pipelines would link us more closely to Iceland, Norway, and Denmark and to Germany and the Low Countries. We, along with our Nordic neighbours would be helping Germany and other less windy and sunny places decarbonise.

I have previously discussed the idea of many large scale energy co-operatives building towards a global network with all of humanity as eventual co-owners and sharing in the profits. I talked about hundreds, or thousands, of projects, many about 4GW. The obvious first place to try such a co-op is in the North Sea, and linked as strongly as possible to several countries, so energy could be sent to wherever the need was greatest and the market price highest. This would be followed by lots of floating wind off the north and west of both Scotland and Ireland. Already some places are emerging as epicentres of innovation based on abundant renewable energy. The Orkney Islands already generate more electricity than they use, and consequently are leading the UK in the speed of adoption of battery electric vehicles. The interconnecting cables taking electricity back to the Scottish mainland are frequently running at capacity. The Orkneys are now exporting green hydrogen back to the mainland and storing it for when the wind is not blowing, to then be turned back into electricity. The European Marine Energy Centre is based in the Orkneys and is at the global forefront of wave and tidal energy research and development, and in a small way deployment.

Throughout this book I have argued that people will want to move to wherever energy is cheapest and opportunities greatest, as we have done throughout the history of our species. For millennia this was to where biological energy, in the form of food was more readily available. During 'The Fossil Fuel Age' it was to coalfields, then to oilfields. Now, as 'The Solar Age' evolves, the movement will be to where various forms of renewable energy are most abundant. By this logic Northern

Scotland could emerge as an epicentre of new technology and new ideas focused on pollution minimization and social solidarity much as is already happening in Northern Sweden. Both areas could well become centres for in-migration. Perhaps both Thurso and Aberdeen might grow, along with Luleå and Skellefteå, into very much bigger towns and cities, maybe housing millions.

Offshore wind would supply the vast bulk of the energy requirements of the British Isles. Our energy needs would of course be very much less, as we travelled less, wasted less and purchased less. All new houses would be built to be well insulated and laid out to maximize solar gain through passive solar space heating, solar water heating and most importantly photovoltaic solar panels. New houses would not be built with chimneys or roof slates or tiles, but simply with modern solar roofs. Most new housing would generate more electricity than it used and heating would be with heat pumps or district heating networks. The older housing stock would largely have gone through a process of eco-retrofitting to make it warmer and cheaper to heat. The vast majority would of course have solar panels added. Cars would mainly be collectively owned by car sharing clubs and re-charged at members' houses.

If one considers the energy economy of Herefordshire, we as a county collectively have been spending a fluctuating amount of money, which over the last couple of decades may have swung as widely as between a quarter and a half billion pounds per year, to include all electricity, heating and transport fuels. This has for decades represented a huge drain on the prosperity of the people of the county. Nearly all this money ends up in the pockets of oil and gas corporations that contribute nothing to the wellbeing of our county.

The changes I have outlined would mean that the energy requirements of the county would have decreased, and a large part of the remaining demand would be generated by rooftop solar panels, and by field scale agrivoltaic co-operatives. The county would of course buy in some energy, but as this would all come from offshore wind co-operatives (of which we would all be members and sharing in any profits) the drain on the economy of the county would be reduced to virtually zero. This would add to the general air of prosperity while reducing consumption. We would all have more time to enjoy the simple pleasures of life, like walking in the countryside and eating good food, which brings me on to the next topic. We all in Herefordshire, the British Isles and indeed the whole world could be living and eating more healthily.

Farming: Herefordshire and the British Isles

I want us to try to envisage what the farmland of Herefordshire and the Welsh Marches might look like in the early 2030s once the impact of the changes that I am advocating have really begun to take effect. In Chapter Four we proposed a global network of about 39,000 model farms, roughly one for each of the planet's communities of about 200,000 people. We envisaged these model farms would have substantial investment of about 25 to 30 million pounds per annum coming from global taxation, totalling an annual investment of about one trillion pounds. With this level of funding our global network of model farms would be recruiting millions of students and apprentices studying, developing and applying new skills and techniques of land management. Some of our students would travel, spending many months or years working on projects where new methods were being developed and then bringing those methods back to their home farms. Wageningen University and Research centre in Holland would play a key role in developing parts of the farming syllabus, and so too would any number of farming pioneers like the ones we discussed in Chapter Four.

The Herefordshire model farm would buy or rent a considerable area of land, and over the first few years of operation this might extend across many thousands of hectares. It would also seek to come into partnerships with a number of existing farmers who want to change or develop their businesses in more sustainable directions.

Herefordshire has overall some very good soils, but many have been abused by years of over ploughing and too heavy applications of chemicals. The organic matter content of the soil has diminished, which combined with excessive ploughing, especially on sloping land, has led to soil erosion. After rain our rivers run red. When they are not red through soil erosion they are often green through eutrophication and algal blooms caused by nitrate and phosphate pollution. The Wye and Lugg are in crisis now. [124] By 2030 we could restore them to their rightful splendour.

We have previously looked at the banning of some chemical inputs and the taxing of others. Ploughing and excessive cultivation will have been discouraged. Livestock numbers will have decreased and intensive poultry houses will have been banned. Hedgerows and bank side ecosystems will have been re-established. Water companies will have been taken back into public ownership under locally controlled municipal structures and regulatory bodies such as the Environment Agency will have been strengthened. All this will have helped our rivers flourish once again. Our model farm will buy some of these farms where the soils have been most damaged and it will rejuvenate them, and restore the rivers and stream that flow through them. It will experiment with a wide variety of crops and systems. It will learn from some outstanding examples.

The farming model I am advocating draws on a number of models, each slightly different, but all overlapping. Agroecology, permaculture, regenerative and organic systems are all different, but it seems to me that the very best farms draw inspiration from each of these schools of thought about sustainable land use. They adapt the principles to their own farms, each with their own unique set of challenges and opportunities. There is another set of ideas emerging around the concept of landsparing, and it seems to me that greenhouses provide one of the best aspects of this concentrated production. By looking at half a dozen outstanding farms and drawing inspiration from them all we can begin to envisage what our model farm in Herefordshire might look like, and also see an emerging vision for our global network of farms.

So let us envisage a model farm within the county with various land holdings in different locations. All would have plenty of financial and human resources to put in, but no agricultural chemicals or fossil fuels. Probably they would all be organic and grow an extraordinary variety of crops. Overall they would be seeking to promote the concept of landsparing, or growing more crops off a smaller area and so leaving more space for nature to heal.

A large multi-story greenhouse might be built, like the one that Growing Power planned for Milwaukee which I previously mentioned. It might be located on the Rotherwas Industrial Estate or some other peri-urban setting, maybe on the southern industrial fringes of Leominster. It might be three, four or five stories high, with lift shafts, meeting rooms and food processing rooms on the northern side. Mushroom farming and artificially lit hydroponics could take place on the ground floor, perhaps with aquaculture above that and berries and salad crops on the upper floors. A ground floor space on the sunny south of the building would probably have deep fertile compost and soil growing an experimental orchard of peaches, apricots, physalis, figs, and maybe oranges and lemons, maybe even mangos and bananas.

Bananas have been grown successfully in Iceland using geothermal heat. With renewable energy, heat pumps and high thermal mass it ought to be possible to sustainably keep the greenhouse warm, and with LED lighting it is possible to grow crops in situations where the sunlight is too weak or non-existent. The greenhouse would have some sections super-insulated and kept quite hot for germinating certain seeds or growing tropical crops, but other sections of the greenhouse would be more like any normal greenhouse. It would be advantageous to build it quite large, so the ratio of external surface to enclosed space was low, and the spaces for experimentation were many. It would probably need a footprint of a minimum of ten hectares.

Parts of, or indeed the entire greenhouse might be made of the new clear photovoltaic panels which are currently been trialled in Holland. In which case the greenhouse might not only be able to generate its entire energy requirements but be a net energy exporter. If that was the case it might become a model to be replicated in other cities around the world.

The greenhouse would seek to be growing many hundreds of different crops, from fruit and salad vegetables to fish and mushrooms. Beehives could be located within the greenhouse with the bees having access to the crops within and to the outside ecology.

Out in the Herefordshire countryside, ideally on soils damaged by excessive chemical use and excessive ploughing, another farming experiment would be under way. Here the challenge would be to rapidly build soil organic matter, and so sequester carbon, while simultaneously producing a vast range of excellent quality food crops. Again this would best be done on a fairly large scale, on a farm of several thousand acres. Let us assume that initially the bird and mammal populations were very low. The land could be assessed and those small bits of existing nature could be protected and extended. So hedges could be allowed to grow a bit higher and wider to act as wildlife corridors and as windbreaks and reservoirs of mycorrhizal fungi.

The existing fields would be divided into strips, with lines of fruit, nuts and useful timber trees acting like powerfully productive hedgerows. Strips of land, perhaps eighteen metres wide, would be used to grow a wide range of horticultural and arable crops and for a wide range of livestock to be carefully rotated along these strips. This is the kind of farming that I have described previously being pioneered by George Young of Fobbing in Essex and Wakelyns farm in Suffolk. They each operate with tiny labour inputs. If we had hundreds of highly skilled and enthusiastic people working to develop these kinds of systems all on one farm an amazing range of experimentation could be achieved and a vastly greater range of crops, and combinations of crops, could be grown.

Multi-species livestock farming, modelled to some extent on what White Oak Pastures has achieved, with a zero waste principle, utilizing every part of every animal, and with on-site abattoirs and direct sales to customers. The livestock would have, almost as their primary purpose, the building of soil carbon and fertility. As the farm expanded the livestock would be moved to more degraded land, and the newly fertile land would be increasingly devoted to fruit, vegetables and grain crops, in an ecological succession. The diversity of crops grown would be extraordinary.

The national collection of fruit trees at Brogdale in Kent has four thousand varieties of fruit trees, and these are just the types suitable for outside orchards, like apples and pears. The Eden Project and Kew Garden both grow an extraordinary range of plants within greenhouses. Nowhere is yet experimenting with just how many crops could be usefully grown to feed a local population. That would be at the heart of our global network of model farms. Each of these model farms might be growing several thousand varieties of fruit, vegetable, grain, nut and seed, and many of them would also be producing many other foodstuffs, from meat and dairy to seaweed and samphire.

Part of the farm could be used for an experiment in agrivoltaic farming, modelled on the Colorado Agrivoltaic Learning Centre. The support structures and crop protection provided by the frames supporting the panels could be used to train climbing beans, squash, raspberries or any number of other crops. The panels might also prove useful as shelter for livestock. It will be important not just to plonk down field scale solar and then to think how best to farm the land around these panels. Instead, careful planning of the farming systems one is aiming to develop will determine the design and construction of the solar arrays. Much experimentation will be required to optimize various configurations of solar panels and farming systems.

Just over the border from Herefordshire are the uplands of mid Wales. Much of the land here is currently under ecologically poor and economically unprofitable monocultures. Upland sheep farming and conifer plantations are the two biggest land use systems. If, say, it was decided to decrease sheep numbers by 80%, and also conifer plantations by a similar amount, it would free up a vast amount of land. Some sheep and cattle farming could remain, with low stocking densities of hardy species, where on-farm wildlife can flourish. James Rebanks book 'English Pastoral' beautifully describes his family's farm in the Lake District over the last three generations and how people and wildlife are now thriving there.

Much of Mid-Wales has very high rainfall, is windswept, has poor acidic soils and has generally been considered poor quality farmland. Re-establishing peat bogs on the most suitable areas would have tremendous benefits in terms of carbon sequestration and be good for many rare and endangered species. It would also act to hold back water, so helping reduce the risk of floods and droughts downstream. Wales does have some small vestigial temperate rainforest, especially in some of its wettest and steepest west facing valleys. Expanding these areas of ancient deciduous forest would be an exciting rewilding project. Re-establishing peat bogs and deciduous temperate rainforest might become the dominant changes on the wettest upland areas, but many areas of better soils might well be used for highly productive farming systems, as we have described for Herefordshire.

Parts of Wales, such as the wide fertile valley between Old and New Radnor, used to grow a lot more orchard crops, grains and vegetables. After the coming of the railways cheaper goods could be brought in from elsewhere and the area became mainly used for livestock farming. Under the changes envisaged in my Global Green New Deal there would be taxes on energy alongside investment in local food economies through the network of model farms. Wales could grow a very much greater variety of crops than it does now in these broad fertile valleys.

The big greenhouse at the National Botanic Garden of Wales in Carmarthenshire is one of my favourite new buildings. I think it is beautiful. It is the largest single-span greenhouse in the world. It is partly earth sheltered, being built back into the hillside to protect it from the cold north wind, yet open to the south to maximize solar gain. The earth sheltering will also provide thermal mass, acting as a radiator during the night to help keep it frost free. It shows us the extraordinary range of plants that can be grown. If towns and cities in the colder northern regions, like Thurso and Aberdeen in Scotland, or Luleå and Skellefteå in Sweden, are to grow very much bigger, it would be extremely interesting to see how they could adapt the design concepts of this Carmarthenshire greenhouse in order to produce local food in their challenging climates. They might want to add ideas from the multi-story greenhouse I discussed as a possibility for Herefordshire. One prediction I do make it that in future a very much greater proportion of the global food supply will be grown in greenhouses, especially in colder climates.

Chapter Six

Creating Change: What we can do

Choosing Engagement

Faced with the huge range of crises facing us it is only too easy to feel overwhelmed and powerless. We have had decades of information and advice about climate change, habitat loss and many of the other aspects of our climate and ecological crises. We have had even longer to ponder the seemingly intractable problems of war, poverty, homelessness and a vast array of other aspects of human suffering. Sometimes nothing seems to be getting very much better, and a great deal of things seem to be, and actually are, rapidly becoming very much worse. We are indeed in a precarious and dangerous point in our history as a species.

Much of our response, both as a society and as individuals is to distract ourselves and try not to think about it. Our media and our politicians would rather we focused our thoughts on celebrity trivia, on sport, on shopping and entertainment, on anything that keeps the wheels of consumer capitalism going just that little bit longer. However as the climate crises unfold, floods, fires and extreme weather events are forcing the issue into people's lives. Scientific reports grow ever more concerned and cry out for action. The response of nearly all the media and politicians is utterly inadequate. They focus on any technical innovation that may offer some hope for them to keep going with a generally business as usual approach, as if one single techno-fix has any hope of reversing our onward march toward catastrophe. The simple truth that pretty much everything needs to change seems totally beyond their comprehension.

A culture of denial permeates most of our politicians, our media and many people. While they are in denial of the scale of the problems we face they remain blind to the possibilities of doing things differently, of creating a better system. The focus of such people is often on the distractions of modern life: of building more wealth and indulging in ever greater levels of luxury, as if these things will give them happiness and security.

For decades much of the debate has focused on individual action to address particular often very narrowly defined problems. If we live in a global economy whose very function creates or exacerbates these problems then although we may win the occasional small victory the overall situation will only deteriorate. That is why we need system change. It is the task of humanity to design, create and implement a new kind of system. The time window we have in order to do this is very tight. We may not succeed.

It seems to me humanity is at a tipping point. If we continue with business as usual we will slide ever deeper into this multi-faceted crisis. If, as seems abundantly clear, our established system is incapable of acting to change direction, then it is up to us to create a new system. However none of us can, on our own, change a global political and economic system. It can feel overwhelming . . . impossible . . . hopeless.

I want to make the case that each of us has very much more power than we may think. The situation is very far from hopeless. Our power is vastly variable, depending on our wealth, our platform, our skills and our abilities. The choices we make as to where it is best to put our time, our money and our effort will vary according to what life stage we are at, what resources we have available and what we see as the best possibilities to engage with this bigger picture starting from where we are at now in our own lives, and in our own communities.

The crucial first step is to realize the vast interrelated nature of the crisis we are in. Understanding this will shape our actions. Alone we cannot change the system. Together we can. How many of us it takes and how successful we will be will only be known for sure many years in the future when we can look back and say "That was the year the last coal-fired power station anywhere on Earth closed down" and "That was the year when global forests, fisheries and ecosystems started to recover" and, most crucially, "This is the point atmospheric carbon dioxide levels started to fall". For any of that we will need a historical point of view from sometime in the future. Our dilemma is: "What do we need to do today?" and my answer is "Support those who are advocating for system change, and those who are, in myriad ways, creating it". That is a rather complex thing and needs some explanation. Deeply reflecting on, and acting upon, this notion of assisting in a globally collaborative effort to create system change may cause major changes to your lifestyle. It may be liberating and a path of great joy and fulfilment, but it will be a bumpy ride, beset with failures and disappointments, from which we can learn and move forward.

If the first step is to realize the vast interrelated nature of the crisis we are in then the second step must be to realize the vast interrelated nature of the solutions. Emotionally and intellectually this is a very different process from that first stage of looking at the bewildering range of problems in the world and how they are interrelated. Once we start seriously investigating the vast and extraordinary range of better ways of doing things, and how they are interrelated, we start to feel inspired to find out ever more, and to engage and to assist in whatever way we can. This is a joyful journey. One that is immensely energizing and full of hope. It makes us want to engage ever more deeply in this exciting unfolding story of healing our societies and our planet.

No one person can know about all of these amazing and positive ideas, technologies, projects and developments. They are fragments of an emergent system. There is no master plan. This book is my personal sketch. Many other people will have their own versions, their concepts of a better, fairer, less polluted, more peaceful future. That is fine. Epochal shifts never have master plans. They result from the interaction of many changes, from technologies to ideas and beliefs. Elements of change will emerge in one place and move to another, changing patterns of belief and behaviour as they go.

As I write this in early September 2021 I look around me and see people active in many diverse ways in many places all over the world, all part of what appears to me an emerging ecosystem of change-makers. Some are busy protesting with Extinction Rebellion and the School

Strikes movement; some are engaged in complex legal challenges, tying to hold governments accountable for their corruption and dereliction of duty. Some are busy setting up tree nurseries and leading workshops on ecological farming. Some are busy developing technologies like new forms of solar powered desalination, better applications of hydrogen fuel cells or how best to improve electricity grids as we shift over to renewable forms of energy. I am busy here bashing away at my keyboard, trying to connect up the dots. My task is to help myself and my readers understand the interrelationships of these many aspects of our unfolding possibilities as we try our hardest to secure the best future for our species that we can, given the predicament that we are undoubtedly in. I wish more people would be more alive to the possibilities to do things differently, and better. It is a process of exploration overflowing with joy and wonder.

Individually we have choices. I see why many people choose the path of denial; it is a way to protect themselves from the enormity of the problems facing us all and to stay safe in their familiar comfort zone. I think I can assume that anyone reading this book is not really in that psychological space. I assume that the vast majority of my readers are horrified by the damage and are to some extent active in trying to help humanity change course. The one clear message I have to you all is to engage with the many possibilities of positive change and especially how they are all interrelated. For all the great basket of problems facing humanity there is an equally overflowing basket of possible solutions. I have used this book to draw attention to some of them, and how they could become the tools of system change.

Daily we are witnessing the epochal crises of the dying days of 'The Fossil Fuel Age'. The better ways of doing things that I have been advocating may the basis of system change that may usher in a new era, which may in future be known as 'The Solar Age'. Only time will tell.

So, assuming my readers are all keen to help reduce human suffering and planetary damage, and are keen to help create a better system, the question is how best to engage. What is the best use of our time and energy? My first bit of advice is to focus on the really big things. Look for ideas, policies and practical applications of things that make the biggest and most positive difference.

Engaging with Climate Justice

Many of us are concerned about what is the most effective action we can take to make our lives more sustainable and ethical, and how we can most effectively reduce the amount of carbon dioxide emissions that we are responsible for. Allegra Stratton, who acted as the prime minister's climate change aide, recently advocated a number of micro steps that individuals can make, such as not rinsing dishes before putting them in the dishwasher.[125] This is part of a long tradition of decades of advice stressing the tiny steps that individuals can make. It is a totally hopeless path, given the scale of the problems we face. Such advice is dangerous in that it distracts attention away from the bigger and more difficult actions that absolutely have to be made. And that is precisely the point of this kind of advice: the established power-holders want at all costs to prevent people making the big changes, which inevitably mean a transfer of wealth and power away from where it currently resides.

These types of tiny domestic actions might if cumulatively applied over the course of a year save a gram or so of carbon emissions per household. There are 27.8 million households in the UK, and so if they each made carbon savings of one gram this would result in a net saving of 27.8 million grams, or 27.8 tonnes. So if every household in the UK took such actions we could save 27.8 tonnes of carbon emissions per year. Meanwhile the joy rides into space recently undertaken by Jeff Bezos, Richard Branson and others emitted very much more than this per minute, per individual.[126]

The work of the American economic anthropologists Richard Wilk and Beatriz Barros looked in some detail at the carbon footprints of a number of billionaires and found that they were all responsible for individual carbon emissions of well over a thousand tonnes per year.[127] As these people compete to have ever bigger houses, super-yachts, more private jet planes and now joy rides into space their carbon footprints are spiralling ever upward. Wilk and Barros estimate Roman

Abramovich's personal carbon emissions are at least 33,859 tonnes. If every single one of us alive on Earth (all 7.8 billion of us!) made the kinds of lifestyle changes advocated by Allegra Stratton it would take us 4 years and 4 months to equal the emissions that Roman Abramovich alone makes every year. There are currently 2,095 billionaires in the world; their number is growing as is their individual and cumulative wealth and their individual and cumulative carbon emissions. And this kind of calculation implies that poor people in poor countries ought to reduce their already miniscule emissions. It is a nonsensical approach.

The average carbon emissions per capita globally now stand at about 5 tonnes, and we have to reduce this figure to zero as quickly as possible. The average emissions of a person living in rural Africa might be about one kilogramme, whereas, as we have seen, the average emissions of the very wealthy are over a million times higher. This shows where the scale of change is required. If we are to reduce total emissions the big gains are to be made in focusing the stringent measures on curbing the emissions on the very rich. The poor African might then be assisted in lowering their emissions while improving their quality of life, which is of course a worthwhile goal in itself, but insignificant in terms of reducing overall global emissions.

It seems abundantly clear that we need to free the world from the damage that billionaire lifestyles inevitably create. So it might be of use to make the kinds of tiny changes advocated by Allegra Stratton, and countless others before her, but it is absolutely vital to also eradicate the billionaire lifestyle from the planet, and to do that through changes to systems of taxation, wealth accumulation and much else. The politicians elected under established systems have palpably failed to lower emissions and will continue to do so as long as they fail to address the excessive emissions of the very richest people in the world.

The carbon emissions of billionaires are just one aspect of the complex web of problems facing humanity, but it is indicative of the kinds of changes we need to make to create a very much more sustainable path for humanity to progress along. It is one example of how our current systems of politics, economics and the rest are failing us, and that is why we speak out, demanding system change.

Climate justice is a key concept in system change. It is the poorest people in the poorest countries who have the lowest emissions yet are most likely to die from droughts and famines induced and amplified by climate change. In the UK, and elsewhere, it is the poorest people, who often are people of colour, who live in the areas with the worst air quality. As we have seen it is the excessively rich, who are disproportionally white, who have by far the highest personal carbon emissions. They also have most options to avoid the risks associated with extreme weather events, as they can more easily move house, afford air conditioning and insurance cover and have so many other advantages.

If we imagine a billionaire with annual emissions of many thousands of tonnes and an African subsistence farmer with emissions of say one kilogram and think about a new kind of system that allowed each human on Earth an equal carbon budget of say a few hundred grams, then what changes to their lifestyles would that imply? For the African farmer it would be very little, say the replacement of their paraffin hurricane lamp with a small solar panel and a few LED lights, or her weekly bus to market changing from diesel to electric. Every single aspect of the life of the billionaire would need to change. His private jet or indeed any air travel would go; the super-yacht and his fleet of cars would all have to go, so too most of his houses, many of his gadgets and possessions and even probably some of his clothes. He might indeed be able to have a comfortable lifestyle within a strict carbon budget. It might even be a happier life, less stressful and more deeply embedded in community, but it would be a life very unlike the one he leads now.

Those who are campaigning for system change all share a concern for climate justice, and that is something no political party anywhere yet has effective plans to deliver, and that is why the actions of the School Strikes movement and Extinction Rebellion are so vital. They are helping to shift the Overton Window, to broaden the possibilities of what politicians can then make into policies.

*

Individual Action and System Change

We have had decades of advice about how to shop more ethically, how to insulate and draught proof our homes, advice about cutting down our emissions by choosing how we travel and what we eat. Meanwhile the ecological and climate crisis has become rapidly more terrifying. Of course those of us who really care have made some changes in some areas of our lives. No doubt there is more we could all do, but to focus on this is to miss the point.

The biggest carbon emitters and polluters in general are little concerned with how those of us who do care are changing our lifestyles. If those doing the vast majority of the damage are very wealthy individuals who simply don't care, or are corporations intent on maximizing their profits irrespective of the social and ecological consequences, or those industries whose very existence is predicated upon pollution, or the military and political processes that are obsessed with power, dominance and control without respect to the wider damage done, then what are we to do?

We absolutely have to change our systems of economics, of politics, of the whole technological basis of the global economy. We absolutely have to change how we manage our global commons, from the forests and oceans to the air we breathe.

For those of us who do care our individual lifestyle choices are important in that they show to ourselves and to the world who we really are. They help define us as people. These choices do indeed have some impact, but nowhere near enough. The big changes are to be made in other arenas: political, economic, and technological and land use, to name a few.

So, if we want to influence these areas where the biggest gains are to be made we have to realize that it is neither easy nor impossible. These changes will happen with enough people getting engaged and actively advocating for better ways to do things, which usually means replacing the failing systems rather than putting all our energies into just endlessly clearing up the mess created by the failing system. This lesson applies across many fields. Let us start off revisiting two contrasting approaches to homelessness.

In Chapter Two (currently page 79) under the title 'Philanthropy or Taxation' we compared the enormous charitable efforts going into helping the homeless in USA and the lack of homelessness in Finland. In USA, despite huge effort and copious charitable giving, the numbers of homeless people continued to grow. Their situation became ever more precarious due to the nature of the American economy being designed as a conveyor belt to produce homelessness, with low wages, insecure gig economy jobs, private healthcare and many other factors undermining basic economic security. Meanwhile in Finland the 'housing first' policy gives everyone a legal right to a home. Excellent free health and social care, combined with higher wages, greater economic security and more secure housing tenure means the number of people becoming homeless is very, very much lower, and can easily be dealt with through the provision of homes and care packages organized and delivered by well-functioning and well-funded municipal authorities.

If we look at the situation of homelessness in the USA we see how it is self perpetuating. The very people volunteering time, energy and money to help the homeless, vote in elections and have to choose between two political parties that have over recent decades only enacted policies that have made the problem worse. Perhaps some of this voluntary effort would better be refocused on creating structural change? Higher taxes on the rich and on corporations, free health and social care, more secure housing tenure, higher minimum wages and better educational provision for the poor could all be brought in. Politicians like Alexandria Ocasio-Cortez are leading calls for a Green New Deal. They are doing much organization of charitable works in the here and now, for example to help victims of hurricane Ida, but also and critically working to create the bigger, more permanent structural changes that would make American society very much fairer, and where a generally more Nordic model of better social provision would mean that problems like homelessness could be effectively eliminated.

In the UK we made massive social gains through the creation of the NHS and the welfare state under the Attlee government in the post-war years. These gains have been severely undermined over recent decades, and now under the truly ghastly government of Boris Johnson the NHS is in grave danger of full privatization and conversion to an American-type system of healthcare. It would be structural change in an utterly socially regressive direction.

The whole focus of this book is to advocate for huge structural change in the opposite, positive and socially just and ecologically regenerative direction. The whole global movement for People and Planet that I have described is pressing for these kinds of changes in countries all around the world.

This book is my attempt to advocate for such a path on a global scale. We are a long way from achieving most of the kinds of things that I am advocating, such as a global legal right to physical and economic security, a home, free health and social care, a legal right to live with clean air and water, healthy ecosystems and a stable climate. In practical terms change will happen in a cascade of interlinked changes.

Activism and Advocacy for System Change (One week in September 2021)

COP26 was held in Glasgow in November 2021. In the run-up to this event, climate activists were getting mobilized in many ways, trying to exert pressure on governments. All manner of groups were active in all sorts of ways. Some were working to lobby within the corridors of power; more were outside that loop, out on the streets and active within communities around the world. Governments tried to exclude those pressing for system change and climate justice from these talks as their message so directly threatens the whole economic and political worldview of nearly all governments.

Fridays for Future is main organization within the school strikes movement and co-ordinates many global actions. They organized another impressive global climate strike on 24th September 2021, which brought hundreds of thousands of activists onto the streets of cities all around the world, with the biggest one in Berlin, just days before the German Federal Elections. Their message is a powerful call for climate justice. The Fridays for Future website has one of the clearest calls for system change and climate justice that I have come across and I will now quote it at length:

The climate crisis does not exist in a vacuum. Other socio-economic crises such as racism, sexism, ableism, class inequality, and more amplify the climate crisis and vice versa. It is not just a single issue, our different struggles and liberations are connected and tied to each other. We are united in our fight for climate justice, but we must also acknowledge that we do not experience the same problems; nor do we experience them to the same extent.

MAPA (Most Affected Peoples and Areas) are experiencing the worst impacts of the climate crisis and are unable to adapt to it. This is because of the elite in the Global North who have caused the destruction of the lands of MAPA through colonialism, imperialism, systemic injustices, and their wanton greed which ultimately caused the warming of the planet. With both the COVID, climate, and every crisis in history, overexploited countries and marginalized sectors of society are systematically left behind to fend for themselves.

The time to join the masses and follow the lead of the environmental defenders and workers has been long overdue. Reparations to MAPA must be paid for the historic injustices of the richest elite, drastic emission cuts in the Global North, vaccine equity, cancellation of debt, and climate finance are only the beginning of these. Together we will fight for a just future where no one is left behind. The historical victories of collective action have proven the need for youth to stand united with the multisectoral, intergenerational struggle for a better future for all; a future where people and planet are prioritized. [128]

In writing this book I hope that the basket of ideas I am proposing chimes with the global network of activists advocating for climate justice. Of course none of us has a blueprint for the future, but many of us share common aspirations about how a better system might look and how we might help bring it about. There are very many strands to the process of change. While Fridays for Future were busy organizing and creating their global day of protest on Friday 24th September 2021 millions of other people, who were not out on the streets with them, were wishing them well, while working away on other aspects of change. Some were busy inventing things, developing the technological basis of a less polluting and circular economy. Others were plotting how to create change in the political sphere by replacing a number of Tory MPs with Green MPs. The range of actions is bewilderingly large and all broadly pushing in the same direction. There are two more examples from the realm of advocacy that I was involved with and which took place in the same week in late September 2021 that Fridays for Future were getting the youth out on the streets.

The COP climate talks due to take place in Glasgow in November were of crucial importance. It was vital that civil society exerted as much pressure for more ambitious carbon reduction goals than nearly all governments are prepared to countenance. Lots of groups were actively engaged in this. One of many interesting and inspiring examples was the Camino to COP26, where many people joined a pilgrimage walk, or walks, with some walking from Europe, the main group walking from London, but joined by others from other locations, all gradually coming together as they walked to Glasgow. The vast majority of people just walked a section for a day or a few days rather than the whole route. I joined the Marches Camino and walked from Hereford to Leominster. We had blessings from the Bishop, along with a Sufi poem and Buddhist chant, as we left Hereford Cathedral, and along the way we received a warm welcome and kind hospitality. The Camino was organized by XR Faith Bridge and their website was full of the most inspiring, inclusive, loving, caring and hopeful language.[129] This spirit was reflected in relationships within the community of walkers and the wider community who came out to support us. I would strongly recommend my readers to join in these kinds of activities. They are powerful elements of change and of personal empowerment. They strengthen and deepen our sense of community. This is a community that embraces our personal embeddedness and connection to the biosphere and to our global human family. I will now quote a paragraph from the Camino to Cop website:

We are united by our faith; a faith that we can advocate and influence and be the change that we want for the world. We choose to walk to COP26 as a practice of that faith, an act of connection with the earth on which we walk and the people with whom we walk and the communities through which we pass; and we make our way in kinship with the people and creatures of the earth who are suffering and displaced by climate and ecological breakdown. We do so peacefully and lawfully, ready to engage and learn, because we care and we have hope....

The other bit of advocacy work I was involved with that week was a talk I gave via Zoom to the Millichap Peace Fund. The few dozen people who signed-up for this session generally came from quite traditional peace, social justice and faith backgrounds. Many of them were relatively unfamiliar with the broader demands of system change. My task was to articulate what is meant by system change, why it matters and how we can assist in bringing it about. My talk was a brief synopsis of this book. I tried to outline my vision of a better future and how we might get there. No mean feat in forty odd minutes. The talk generated a great range of questions.[130]

In the week or so that I gave this talk, and walked for a day with the Camino to COP, and Fridays for Future were organizing their global day of action, millions of others were actively engaged in contributing in myriad ways to creating a better system. I am in awe of so many of them, whether it is inspirational farmers, educators and tree planters like Patricia Kombo with her PaTree initiative in Kenya, or technology geeks like the people at Riversimple developing the most ecologically sustainable car yet designed, or the growing network of Green Party local councillors, they are all playing their part in this hopeful and unfolding narrative of change.

So, perhaps my first bit of advice to my readers is to support this whole global cornucopia of such groups. Some of you are no doubt already active in some these groups and some of you just support them in a token way. There are many other ways to help change the world for the better, and none of us has the time and energy to put into many campaigns and activities simultaneously. We have to choose day by day and week by week what action feels right for us as individuals. The important thing is to engage, to become more of a participant in change-making, and not to be a bystander bemoaning how bad things are and passively going along with it all. By listening to and engaging with these positive change makers, and as far as possible paying a lot less attention to what the tired old voices of the incumbent power structures are saying, we find a path towards greater joy, energy and empowerment.

Grassroots Political Change

For years there have been many worthy reports from The Club of Rome, the United Nations and books by academics advocating a shift in policy direction toward a better, fairer, less polluted future. For any of these top-down attempts at leadership to actually deliver results there has to be active support from national governments, local authorities and also from local communities. Following on from the Rio Earth Summit in 1992 lots of good initiatives were supported by a scattering of national and local governments, and local communities formed Local Agenda 21 groups. As we have previously noted excellent initiatives like the Aalborg Process grew from this. However, generally these top-down initiatives have been ignored by national governments, who are more concerned with an outdated notion of national self-interest, or indeed personal financial interest, rather than the wellbeing of the planet and its people at large. For every step forward there has usually been a step or two backward. There has not been the political will to deliver on these well-intentioned reports and pleading.

The challenges facing humanity need political change at all levels, from the global to the local. We, as individuals and communities, can seek to create change in many ways, but although we would love to instantly rid the world of many of its worst governments, our ability to do so is very limited. We can of course express our support for activists pressing for better governance in, for example, Belarus, but our ability to play an active role is tiny. In our own local communities we can make a difference.

Many people have given up with party political activity as a means to create change. I want to argue that it is a vital element in the processes of change, and it is something we can all easily do within our own communities.

I have been a supporter of the Green Party since the early 1970s, when it was known as the Ecology Party. It seemed to be the party with by far the best policies. We formed a local Herefordshire branch of the party in 1982. Over the years, both nationally and locally, we made very slow progress. Between 1976 and 2018 we gradually increased our number of local councillors in England and Wales from one to 173. Then on one day, Thursday 2nd May 2019, we leapt up to 362, a net gain of 189 seats, more than doubling the total we'd spent the previous forty-three years building up. It is important, and heartening, to realize that political change can be sudden. It is also important to acknowledge that it only happens after much grassroots organization and preparation.

It is at this very local level that grassroots political change takes place, where teams come together and discover their own power and capabilities. It is only from this very small local basis that we can build. The Green Party does not have access to the media or to huge advertising budgets or as many paid staff as other parties. It has to compensate by having bigger and better organized teams of enthusiastic volunteers, keen to engage with people on the doorstep. That can have dramatic results. All it needs to win a local council ward is an excellent candidate, a good-sized team of volunteers and excellent organization. It is also important to have bit of money, but not that much. In recent years the Green Party has become very much better organized across most of the regions of Britain, and is building teams in many diverse communities. In the May 2021 round of local elections the Greens again made really significant gains and now, as of September 2021 have 454 seats across 143 principle authority seats across England and Wales.[132] In local council by-elections over the following months the Greens gained a few more seats. The next round of local elections is to be held on 5th May 2022, when, by my calculations about 6,941 seats are up for grabs across the UK.

People might argue that local councillors don't have much power to change things. In the UK context there is truth in this as local government is so terribly underfunded that it cannot do as much as many of us would wish. Better governed countries in general are very much less centralized and so their local governments are very much more generously funded and able to initiate and deliver a much greater range of services. However these Green gains in local government wards across England and Wales are significant. Good political parties are built up from below. It is only by having excellent and enthusiastic teams of volunteers working at this most local level that we have any chance of replacing some of our worst MPs. Getting more local councillors elected goes hand in hand with gaining vote share at general elections. In Herefordshire in May 2019 the Conservatives lost control of the local council to a very creative mix of Independents and Greens. That massively increased the chances of replacing our two local MPs, who are both Conservatives, with two Green MPs. Again, some might argue that having two more Green MPs will not create all the changes we need to make. Of course this would just be a couple of small pieces in a gigantic jigsaw puzzle of change, but I would argue one of the best. Having Greens in the room, even when they are a minority, allows things to get debated, and sometimes acted upon, which would otherwise just not even be considered. It is a significant way of shifting the Overton Window.

Working at the party-political level should not be seen as in competition with joining the activities of groups like Extinction Rebellion, Fridays for Future or Friends of the Earth, but rather part of the same broad push to create change. Many of our best activists are engaged in many organizations, and there is considerable overlap in membership between, for example, Extinction Rebellion and the Green Party.

Being an activist for Extinction Rebellion or Fridays for Future entails huge goals: system change, falling atmospheric carbon dioxide levels, flourishing ecosystems and radically greater global equality. These are vital goals, but vast and hard to achieve. With such vast goals it is sometimes difficult to know when progress is being made.

Yes, we can feel shifts in public opinion and in media coverage of the big issues of climate change, ecological disasters and the impact that these are having on communities, but actual tangible progress is hard to measure. On the other hand groups like Friends of the Earth have spent years campaigning for the banning of neonicotinoid insecticides, and had partial success in some countries, only to see these gains reversed by successive governments more closely aligned to the agro-industrial farming industries. Meanwhile as they campaign on this issue a thousand other issues are being ignored. We need groups like Pesticide Action Network and Friends of the Earth doing their focused campaigns and we need groups like Extinction Rebellion and Fridays for Future working on the bigger picture, but achieving success in either of these arenas is going to be very much easier wherever we have Green politicians within the decision making process.

While the aims of Fridays for Future are big and nebulous and those of groups like Friends of the Earth are narrow and sometimes seem insignificant in relation to system change, replacing a dreadful Tory MP with an inspirational Green MP is a very clearly definable goal, with a very clear timeframe and broadly impactful across a wide range of policy areas. The UK is due to have its next general election by 2nd May 2024, but it could be at any time before that.[133] The parliamentary seat of North Herefordshire is represented by Bill Wiggin for the Conservative party. At the 2019 election he had a whopping majority of 24,856.[134] It has been considered a safe seat. The Green Party has an excellent candidate in Ellie Chowns who understands the global situation and the kinds of issues discussed in this book in a way that Bill Wiggin never will. If we could replace Bill Wiggin with Ellie Chowns it would be an extraordinary achievement: another piece in the vast jigsaw puzzle of global change. It may not be possible to achieve it in one go. It might take another election to narrow the gap and then one more to win, but it is worth going all out to achieve it at the next attempt. While with groups like Extinction Rebellion it is impossible to say how many activists it takes to win a significant victory, with an election it is much clearer. With a team of a thousand activists we could replace Bill

Wiggin with Ellie Chowns. Being a political activist does not have to be something requiring too much time, expertise or special skills. There are many ways to help; delivering leaflets, canvassing, donating money, making cups of tea or helping organise events. It is all about building effective, organized and coordinated grassroots teams across as many of the political wards as possible.

There is an emotional and psychological element I want to mention. Being part of one of these teams where you can feel the momentum swinging your way is quite an amazing and positive experience. Having spent decades doing odd bits of leafleting and canvassing at a lot of elections where it was obvious that we would not win was quite frankly dispiriting. Then over the last few years the Green Party in Herefordshire has got very much better organized, the team has grown much bigger and we've had some very good candidates. For many years we struggled to get and maintain a single councillor on Herefordshire Council, then over the last five or six years we started winning a number of seats. In May 2019 we won seven seats and together with a larger group of independent councillors replaced the Tory run administration. I played a small part as a member of the team that helped get Ellie elected in Bishops Frome and Cradley ward. That felt very energizing and empowering. Imagine what being a member of the team that gets her elected as one of our local MPs would feel like. It is not often in social and environmental campaigning that one has such a clearly defined goal and such an opportunity to celebrate success.

The Next General Election

The current state of politics in the UK is more polarized and angry than at any time in my life, and probably any time in the last hundred years. The Conservative party has been taken over by extreme market fundamentalists and disaster capitalists from the European Reform Group. The current cabinet has nobody left representing the old one-nation Conservatives. The few remaining one-nation conservatives in Parliament, such as John Major, Ken Clarke or Michael Heseltine are deeply critical of the current cabinet, and many of the better Tories have left parliamentary politics, such as Rory Stewart, Dominic Grieve, David Gauke and Anna Soubry. The Labour opposition looks very weak and deeply divided. The hatred between the Corbynite and Blairite wings is intense, and will not be resolved any time soon. Keir Starmer seems to embody a bland but competent model of business as usual politics. He is neither a unifying nor an inspiring leader. He will do nothing to upset the status quo.

Brexit is proving to be the utter disaster many of us predicted. Many people would love now to re-join the European Union, but that is a distant goal, as quite naturally the EU would be very reluctant to take us back after the chaos that we as a nation have caused to ourselves and our European colleagues over these last few years. The next few years will be a time of unprecedented anger, chaos and confusion in British politics. Thousands of businesses are going bankrupt as supply chains become impossible to manage. Shortages of all sorts from food on the supermarket shelves to shortages of doctors, nurses, lorry drivers, waiters, fruit pickers and many other occupations will continue to create endless crises. Such crises are to disaster capitalists and corrupt government ministers simply opportunities to make money for themselves and their mates, as they have demonstrably proven through their handling of the Covid Pandemic.

The mood amongst the British people is angry. Recent by-elections have seen huge swings. At both the Chesham and Amersham, and the North Shropshire by-elections the Liberal Democrats overturned huge Tory majorities, and won with very comfortable majorities. Meanwhile the Labour party which ought to have won the Old Bexley and Sidcup by-election failed to do so. Over the last few months local council by-elections have seen large gains by the LibDems and the Greens, the Tories losing lots of seats, and interestingly, also Labour losing a few. There seems to be a move toward smaller parties. Plaid Cymru, the SNP and the Alliance Party in Northern Ireland all look relatively strong to me.

Although I would love to see a Green government, the Green Party will not replace the Tory government on its own. Not any time soon.

The damage being done to this country is too extreme and too rapid for any one party to be sure of beating the Tories at the next general election. There is much to be said for working together. Coalitions often bring out the best in all the participating parties. They are the norm in most well-functioning democracies. Listening, co-operating and compromising are powerful political skills. We can learn from each other. We do not need to agree on every last detail. Extreme and bitter divisions exist within both the Conservative and Labour Parties. A coalition formed by the Greens, LibDems, Plaid Cymru, SNP, Green parties from Scotland and Northern Ireland, and also the Alliance Party of Northern Ireland might work very well together. Parts of the Labour party are enthusiastic to join some kind of Progressive Alliance. As the biggest opposition party many Labour members see that they would naturally be leading it. However some other parts of the Labour Party are strongly opposed to even considering co-operation with other parties.

A Progressive Alliance is being worked for in many ways. Caroline Lucas from the Green Party, Layla Moran from the LibDems and Clive Lewis from Labour work together within the Compass campaigning group seeking to build a Progressive Alliance. Informally at some recent by-elections voters have tended to cluster around whichever candidate they saw locally as having the best chance of beating the Tory. Tragically Labour leaders have never supported such an alliance. The Labour Party constitution has a clause requiring them to stand candidates wherever they can. Compass is actively campaigning for Labour to change this. Each of the opposition parties needs to focus their efforts where they have the best chances of winning, and to allow space for other parties to try and win elsewhere.

For any kind of Progressive Alliance to work there will need to be some very basic and simple set of agreed policy objectives to unite around. Committing to some kind of simple ten-point plan like this would be a good place to start. If all the parties could agree to progressing these ten policies within the first one hundred days in office it would be clear to voters what they were voting for. Here are my top ten urgent political goals for any UK Progressive Alliance. You no doubt will have suggestions and improvements, but here is my list as a starting point.

1. Bring in some kind of proportional voting system.
2. Bring in some kind of universal basic income.
3. Invest heavily in a roll-out of renewable energy, home insulation and other policies to reduce carbon emissions and to help ease fuel poverty.
4. Invest heavily in the NHS and reverse all recent creeping health care privatization.
5. Invest more in education, from pre-school to post graduate. Abolish tuition fees and re-introduce student grants.
6. Bring in the Climate and Ecological Emergencies Bill, including Ecocide as a serious crime.
7. Seek to rebuild as close a relationship with the EU as possible, including rejoining.
8. Increase higher rates of income tax, wealth and property taxes.
9. Increase taxes on all forms of pollution, from single use plastic to fossil fuels.
10. Stop all forms of perverse subsidies such as for oil and gas exploration.

These ten points are of course far from an exhaustive list. Having something that is clear and easily understood by all the parties and the general public is an important first step. It is something to build on. No doubt the SNP would want a second independence referendum, and Plaid Cymru might want something similar. All that would be up for negotiation. We need to talk.

It seems to me it would be beneficial for both the Labour Party and the Conservative Party to split up into several parties. I think Tories such as Dominic Grieve and Rory Stewart might have a useful role in creating a more reasonable government. There are many people within the Labour party who would love to work in this way, but their own party hierarchy has effectively blocked any such collaboration, be it under Blaire, Brown, Corbyn or Starmer. British politics remains the confused and dysfunctional mess it has long since been. Given the historical point we are at, both as a nation and as a species, this is a tragedy. The UK is unlikely to offer any kind of global leadership, although many of us would wish it otherwise. However, we might yet be successful. Politics is in a highly volatile state. Things can change quickly. A progressive alliance may emerge, even against the wishes of the Labour hierarchy.

I will keep doing all I can to help the Green Party locally to me. I also celebrate the victories of LibDems and others. The left and centre has massive support on the ground. North Shropshire has elected right wing Tories as MPs and local councillors for years. In the recent by-election the combined vote for LibDems, Labour and Greens was 61.5% of the vote. If North Shropshire is so winnable, any seat in the UK is. Let's work together to get rid of as many of the ERG (European Reform Group) inspired Tories as possible, and that means all the cabinet and most of the serving Tory MPs. If Dominic Grieve or Anna Soubry wanted to stand with a Progressive Alliance against incumbent right-wing Tories I would welcome them. If Labour politicians wanted to join I would welcome them, but at the moment leadership is coming from the smaller parties, including the Greens, LibDems, Plaid Cymru and SNP, from pressure groups outside Parliament and from a few dissident voices within the Labour party.

Lifestyle

Having stressed the importance of engaging with the larger processes of structural change, I don't want to overlook the importance of our individual lifestyle choices. They do matter, and in more ways than we might be aware of. These choices help shape us as people. They are the clearest outward manifestation of our most personal values, beliefs and aspirations. They are both crucial to our individual development as human beings, and to how we are perceived by others. They can help us build networks of solidarity and community into which we fit with a growing sense of belonging. While we might be critical of some aspects of this broader movement at times, it is important to understand that what unites us is very much greater than what divides us. It is also useful to keep appraising our own actions. Sometimes we will be involved in efforts that don't succeed, or on hindsight might seem a little misguided. It is all part of our individual and collective learning. We are all caught up in the crazy and chaotic unfolding of a newly emergent system which is everywhere in conflict with existing, incumbent and dominant systems.

So, our individual lifestyle choices have importance to us personally, and if taken up by many millions of people, can have powerful direct impacts on carbon emissions, pollution in general and on social justice. There are two aspects of this I would like to investigate. One is the practice of avoiding those choices that cause the most damage, and the other is actively seeking out the choices that promote the broadest range of social and environmental benefits.

The first aspect, the one of avoidance, has perhaps traditionally got more attention. So during the Apartheid era many people boycotted South African goods. More recently many people have been boycotting Israeli goods and services because of their mistreatment of Palestinians and others have boycotted Barclays Bank because of their funding of fossil fuels. All this can be of use. The problem is that there are so many countries and corporations that we could justifiably boycott that it becomes dreadfully time-consuming to avoid them all, and of questionable benefit if it is a few people boycotting something which is being purchased by many millions of people. Also if we switch our purchasing power to a rival product we often find out that that it is not much better. So if we boycott one bank, supermarket or country only to switch to a slightly less awful but similar one, then our action does not amount to much.

More useful is to reject the worst things and proactively switch to something very different. In fact the psychological focus of our attention is often best placed on discovering and adopting the very best things and simply avoiding the worst without making a big deal of that side of the equation. Many people are now refusing ever to fly in an aeroplane. This is one of the most effective ways to reduce our personal carbon footprints. If we then spend our fortnight's holiday off on a cycle tour, or participating in one of Extinction Rebellions actions, rather than flying off to the Mediterranean we might well find the experience more rewarding in very many ways. The same applies to supermarkets and shopping. So while some supermarkets are worse than others they all share many damaging practices. Much better, as far as we can, to switch our purchasing power to directly support small, local and ethical businesses or to grow our own food, or initiate buying co-ops with our friends and neighbours, or buy directly from the farmer.

Let us look at two very different things, cars and loaves of bread, and see how adopting this principle of avoiding the worst, while seeking out the best alternatives that we can find, can have multiple knock-on effects that spread though our lives and through society. Of course I write as an individual, and will relate these stories through the changes I have made living my life here in Herefordshire. The possibilities will inevitably be different where you live, but some of the principles will be transferrable.

Cars and Me

My relationship with cars has long been a bit fraught. As a teenager in the early 1970s I was aware of all the negative aspects we are aware of today: their role in climate change, local air pollution, accidents, congestion, dividing communities and making many areas deeply unpleasant to live in. All these things were much debated in the 1960s. In 1970 when I sat my English 'O' level one of the essay questions was simply 'Should cars be banned?'. We were all familiar with the pros and cons. I argued that on balance they should. For me then, as now, their carbon emissions were a major factor.

We had doorstep milk deliveries from an electric milkfloat, as most people did in those days in suburban London where I lived. I had read about generating electricity from the renewables, but beyond hydro this was still more an idea than an industry. We wondered how long it would be before we had electric cars running on renewable energy. This would be better than fossil fuel cars, but not alleviate congestion, accidents or the ruining of urban life with great roads carving through neighbourhoods.

When I got to eighteen I had to choose, whether to learn to drive or not? I did not want to be responsible for more pollution, but neither did I want to be deskilled in relation to everybody else. My choice was to learn to drive, but not to own a car. Then many years later, after I'd bought an old ruined cottage and was struggling, transporting sacks of cement by moped, I did eventually succumb and bought a van. Over the following thirty years sometimes I owned a car and sometimes I didn't. When I was living in cities I never felt the need for one, whereas living in the countryside and doing building work I did. I had a succession of old bangers.

Then about 2008 or 09, once I'd moved into Hereford and married my wife Colette, we started talking to neighbours about forming a community run car sharing club. After a long period of talking, and trying to establish a large enough group to make a club viable, we launched in June 2012. We had lots of help from Robin Coates from Colwall car share club, which was already up and running. Our car sharing club in Hereford has been running for a nearly a decade now, and it is working very well. We share the ownership, responsibility and use of five cars between about thirty-five to forty households. For most of us it is much cheaper than owning one's own car, and has many other advantages. As we have five cars there is choice: a small electric car is best for local journeys, but a big estate is useful for that odd journey where we need to carry a lot of stuff. For most of us our car use has declined and we walk, cycle and use public transport more than we did when we owned our own private cars. For my wife and me some months we don't use a car at all, other months it might be for a few hours, and sometimes for a few days in a month, all very much less than we did back in the days when we owned a car. The car club has helped in some small way to reduce congestion, carbon emissions, and local air pollution while giving us all greater flexibility and has to some extent helped us get to know some of our neighbours in a way we never would have without the club.

In the chapter on Energy and Infrastructure we discussed the big changes required to move from a wasteful linear economy to a radically less polluting circular economy. One key aspect of this is to reduce the number of things we each individually purchase, use and then throw away. For some people giving up one's car and walking, cycling and using public transport is an easy change. For others it is almost impossible. It all depends on where we live and work, what kind of work we do and also our physical abilities and disabilities. There are very many people who would benefit from being members of a car sharing scheme. There are many models from the more expensive commercial ones operating in many big cities to the much cheaper community run ones that do require more effort and organisation from within the community. Getting rid of a fossil fuelled car and replacing it with an electric one is good. So too is replacing individually owned vehicles with shared use ones.

In our car sharing club we set out with the goal of both reducing total cars owned within our community and gradually moving towards more ecologically sustainable vehicles as and when we could afford to do so. We started out with whatever cars our members were prepared to donate, which naturally were a collection of older fossil fuelled cars. We were lucky to then secure grant funding to help us buy somewhat newer and more energy efficient models, but still fossil fuelled. Having a politically sympathetic council helps. Now we have a Green and Independent coalition running the council they are putting resources into helping four or five new community car sharing schemes to get established and to help purchase some electric vehicles. This is of course alongside investing in walking, cycling and public transport and cancelling a by-pass that the Conservative administration had been yearning to build for decades.

Our car sharing club currently has three diesel cars, one petrol and one battery electric. We have always aspired to gradually move from older fossil fuel cars to better hydrogen fuel-cell and battery electric ones. We bought our first battery electric one a year or two back, and as one of our old diesel cars is approaching a point where it is going to be too unreliable to be a useful part of our fleet we would like to replace it with another battery electric car. We are on the waiting list to trial the hydrogen fuel-cell Riversimple Rasa, with the aspiration to lease one from Riversimple. The Rasa is in my opinion probably the most ecologically sustainable car about to enter production. Another car that I have been watching the development of with great anticipation is the Sono Motors Sion. A couple of German teenagers had the idea of building a battery electric car and covering it with solar cells so that it would charge itself from the sun. They built a basic prototype, and from this has grown a company that has designed a really exciting looking car.[135] I think it would make a great addition to our car club's fleet. It is not due to be in commercial production until 2023, which might give our club time to save up to buy one. Our community car share club is financially viable, due to the hard work of a few volunteers including a very competent treasurer. This structure gives us the possibility to gradually move toward cars that are very much less polluting, and sharing them between a lot of us massively reduces each of our individual costs and carbon footprints. It is a model I'd recommend to others.

A Loaf of Bread

Nearly all of us eat bread on a daily basis. The various systems of growing grain impact biodiversity, carbon emissions and our own health. Most of the bread sold in the UK and many other countries is wheat grown in chemically intensive monocultural fields using so called Green Revolution improved seed, which is designed to be used with more chemical inputs. This wheat is milled, often over-refining it and removing the bran and wheat-germ, depleting the nutritional content. This flour is then turned into bread using the Chorleywood process. Writers such as Andrew Whitley and Nigel Slater have long pointed out the damaging effects of this.[136] Over the last few decades there has been a bit of a resurgence in people baking their own bread, and more recently in sourdough baking. Such slow fermentation methods of baking produce very much healthier bread, especially where nutrient-rich older varieties of wheat and other grains are grown, and grown on farms run on sound organic principles and where the whole grain is ground to make the flour.[137] The choice of such high quality, nutrient-dense flours suitable for home bread baking has grown over recent years as pioneering ecologically regenerative farmers have worked to develop high quality grains, often from ancient varieties of grain. Martin Wolfe and a team from the Organic Research Centre pioneered such methods from 2001 at Wakelyns farm in Suffolk.[138] Hodmedod's[139] has developed as a company specializing in the marketing of these wonderful British grown nutrient-dense grains and pulses.

George Young from Fobbing in Essex is another of this pioneering band of people growing and marketing a great variety of nutritious grains in the most extraordinary and productive agro-ecological way. George Young, like the team at Wakelyns, grows and sells a huge diversity of crops, as all good farmers utilizing polycultural and agro-ecological methods might be expected to do. These are exactly the kinds of farmers I would like to see helped under some system like my envisaged Global Green New Deal. By taxing chemical inputs and subsidizing the creation of many millions of long-term, high-quality apprenticeships as outlined under my proposed Global Trust for People and Planet, we could aid such pioneering farmers, and help replicate these kinds of ecologically restorative systems of farming. Combining this with very strongly egalitarian social and economic policies, massive reduction in pollution and all the other ideas outlined in this book has vast potential to improve human health and thus reduce illness and suffering, which would have the added benefit of saving money to be invested in yet more improvements in all areas of life.

Now let us return from these big picture ideas to the things we can each realistically do, practically in our own lives, wherever we live. What we buy matters. Such purchasing decisions directly influence what we eat and therefore our own health. Every single item we purchase influences the global economy in some tiny way. We put our money into something; that shapes what is produced, and how the product is made, and who benefits. Let us look at how this works in terms of the bread I eat.

Over recent years the quality of bread I eat has improved. Partly this has happened as my wife has become very keen on baking sourdough bread and using the increasingly available better-quality grains. We don't bake all our own bread. We buy about half of it, and that mainly comes from an excellent local artisan bakery in Hereford called Nizi. David Nizi sells a wonderful range of bread and pastries from his tiny bakery in Hereford. One of the loaves that we love most is his Black Mountain bread. This is made from YQ, or yield quality flour, grown on Great Bettws Farm near Abergavenny on the Welsh border. Ever since the Covid pandemic started we have put in an order about once a month for a few loaves of bread to be delivered. The deliveries are made around Hereford by cargo bikes operated by Pedicargo.[140] Pedicargo was set up by Will Vaughan over a decade ago, and has grown from strength to strength, with ever more innovative bike and bike trailer designs. By ordering our bread from Nizi, with flour from outstandingly good farms such as Great Bettws Farms' local landrace wheat, and for it to be delivered by Pedicargo we are simultaneously supporting three tiny but excellent local businesses.

So, if we think about how these small life-style choices of cars and bread are influencing our health, our local economy and the wider global context we can see that they both embody some of the same real-world choices. We put some effort into actively seeking out better ways to do things. Some things, like setting up a car sharing club as an alternative to owning our own car required helping to build a community venture, which was not without the efforts of many people, but the benefits, I, and I think most of the members, would say vastly outweigh those efforts. With the bread we have chosen to support three tiny and sustainable businesses, and my wife has invested the time and effort in baking bread, but the quality of the bread we eat has improved and our money is going to people we wish to benefit, rather than to some corporate giant of a supermarket chain. We do still shop in supermarkets from time to time, but for those things where we can easily access better products and services elsewhere, we do usually try and support those better alternatives. None of this makes a huge difference in the global scheme of things if it is just a few people making these kinds of changes, but where many millions of people do make these kinds of changes big shifts in the economy can result.

For many millions of people joining a community-run car sharing club would be great way to give up their individual ownership of a car. By getting Greens and other sympathetic councillors elected the possibilities for local governments to assist in the creation of such groups grows greater. Likewise, with the right political and economic changes, the roll out of better-quality food and farming systems could be developed and many tiny, ethical and sustainable businesses could be helped to establish themselves and so replace some of the worst aspects of the global food system. Individual life-style choices, political activism and system change are of course all interrelated in complex ways. Creating change in one area can be greatly assisted in changes elsewhere. It is a web of change, and quite how and when we make efforts in any one area is often determined by chance conversations and shared ideas and aspirations.

Chapter 7

Postscript: Putin & Possibilities

Introduction: Three Themes

I started writing this book, or at least this version, in September 2020 and sent the manuscript to Glenn, my editor, on 21st January 2022. He returned the manuscript to me on Thursday, 24th February; the same day Putin's armies invaded Ukraine. I then spent the next six weeks divided between correcting hundreds of little spelling and grammatical errors and avidly following events in Ukraine.

It is now early April 2022 and much has changed since 21st January. There are three themes I want to explore in this postscript.

Firstly, the horrific war in the Ukraine. There are some excellent academics, journalists and activists from the eastern parts of Europe who have pertinent insights on the situation in Ukraine and Russia, and on the future of that region.

Secondly, the UK is in an extraordinary mess. Government corruption, Brexit and the cost-of-living crisis are deeply interlinked. Real material poverty is re-emerging at a speed and severity that surprises many people.

Thirdly, the latest IPCC report and recent temperature anomalies in both the Arctic and Antarctica point to a terrifying future, if we do not reduce emissions very much faster than has so far been achieved.

These three themes are themselves deeply intertwined. The path to a better, more peaceful future necessitates addressing them all together. As throughout this book I want to stress the importance of a basket of interlinked solutions. The situation in Russia and Ukraine is critically important, and how we respond to it will to a large extent determine the future for all humanity, maybe for all time. Amidst all the genocidal brutality, ghastly suffering and heroic defence there is also a window of possibility to a very different future for Ukraine, for Russia and for the world.

As I have argued throughout this book the need to stop burning fossil fuels could not be greater. We absolutely have to reduce emissions very much faster than has been the case. The latest IPCC report makes this abundantly clear. Putin's Russia is a hollowed-out economy, utterly dependent on exporting fossil fuels in order to fund its vast military machine. We need to stop buying Russian oil, gas and coal. This embargo ideally would be imposed immediately by all countries. Some countries are ramping up alternative supplies, from Saudi Arabia and other oil and gas exporting nations, as a way of quitting Russian supplies. This does nothing to reduce carbon emissions, and supporting Saudi aggression in Yemen is not a lot better than supporting Russian aggression in Ukraine. Some in the UK are advocating fracking and no doubt the Canadian tar sands industry is working at full tilt. Ramping up these most polluting industries makes no sense. Others are advocating building more nuclear power stations, but this is a very slow and expensive pathway, utterly unsuitable for the rapidity of action now required. There are better options, namely a massive push to rapidly reduce demand and also to ramp up renewables, both of which can be done quickly, safely, cheaply and which can be implemented in such a way as to achieve other beneficial outcomes.

I will return briefly to the energy transition in the final paragraphs of this book. First we need to delve into the situation in Russia and Ukraine and examine the possible ramifications for the region and the world. Then we will look at the extraordinary mess that the UK is in and propose a few immediate actions that could be taken.

War and Peace: Russia and Ukraine
In the chapter on politics I pointed out how Russia has remained autocratic, brutal and over-centralized, and has exhibited a continuity of foreign policy for hundreds of years.[141] Be it under the Tsars, Communists or Putin, Russia has repeatedly invaded and sought to dominate its neighbours. (Page 56) Over these hundreds of years

there have been one or two short periods when a more peaceful, kind and democratic system seemed possible, notably in the Gorbachev years. Such hopeful signs were always brief interludes before a return to authoritarian rule. This authoritarian rule was sometimes dull and relatively uneventful, as in the Brezhnev years, and at other times psychopathic, cruel and genocidal, as under Stalin, and increasingly so, under Putin.

In Chapter Two I discussed pathocratic systems of governance. Putin is exhibiting all the classic signs. The genocide in Bucha comes as no surprise. Putin is behaving in ways ever so reminiscent of Hitler and Stalin. As he personally becomes ever more powerful he is increasingly more distrustful of those around him, ever more paranoid and isolated from reality and unfeeling towards human suffering. The state media is full of lies bolstering up Putin's regime, but making it ever more detached from reality. There has grown a sense of ethnic Russian exceptionalism very like the German sense of their own racial superiority in the Nazi era. Hatred toward Ukrainians, Finns, Poles and the many ethnic minorities within the lands of the former Russian empire permeates the Russian media. As under the Nazis this fostering of extreme racial hatred is a fundamental part of the policy agenda. Much of the Russian population now, as in Nazi era Germany, are hyped-up on hatred and their own sense of racial superiority. We in the West are not immune: populist politicians everywhere seek to whip up this destructive force of racial hatred whenever it suits their own interests. Farage, Le Pen, Orban, Trump and countless others have expressed admiration for Putin. Given sufficient power they might each unleash such terrors. However, none of them has the power, or the lack of checks and balances, that Putin has.

Nearly all Western commentators misread the situation in Russia and Ukraine, especially in the first days of the war. Mostly their reporting had been informed by perspectives from international relations of 'realism' and the strength of 'great powers'. They saw Russia had more weapons, manpower and resources than Ukraine, and therefore predicted Russian victory, or some kind of ceasefire whereby Ukraine would cede territory to Russia. This overly sympathetic view of Russia has also been informed by decades of Russian interference in the British economy and politics. Russian oligarchs have bought power and influence, corrupting our politics and our economy. Brexit was from the start a Russian policy objective. The Vote Leave cabal now dominating our cabinet have deep Russian connections. Much of our media is owned by Russian oligarchs or billionaires with similar worldviews. Much of the left of British politics is so deeply critical of NATO that they fail to acknowledge situations of deeper danger, especially when they stem from Russia. Much of the BBC, Channel Four and other media is influenced by these weak and ill-informed debates among British politicians and broadcasters, whose understanding of Russia tends to be shallow and poorly informed.

Another narrative has been unfolding that is more influenced by a deeper understanding of Russian history, and Russian relations with its neighbouring lands. Over these last few months I have been following this other narrative with a passion. I am very grateful to a number of academics, journalists, politicians and activists from across Ukraine, the Baltic Republics, Poland and Finland and also, and perhaps most importantly, from some people of Russian origin who now live abroad and are Putin's fiercest and best-informed critics.

This other narrative stems largely from countries that have been repeatedly invaded and subjugated by Russia. Ukraine, Poland, Finland and the Baltic Republics and the Czech and Slovak nations, (along with many others) have all been invaded by Russia, some on multiple occasions, yet none of them have invaded Russia. There is a power asymmetry. Their history of being the victims of Russian aggression informs their understanding and actions today. There are voices from these smaller countries that it is important to listen to today.

This other narrative also draws on the importance of networks, non-state actors and morale, rather than simply the 'realist' perspective, as being critical factors determining victory or defeat. From the Russian build-up of troops to the 24th February invasion, and over the weeks since, they have been predicting the likelihood of Russian defeat. Their analysis has proved more useful, and better at predicting the events unfolding in Ukraine, and the ramifications for the world than has that of the realists who have dominated Western discourse. I posted a blog on 6th March predicting Russian failure, and that for Putin this would probably mean either death or the International Criminal Court in The Hague. Such an outcome is by no means guaranteed and Putin's demise might be slow or quick in coming.

If Putin is successful in his war in Ukraine he is very likely to then invade the Baltic Republics, then probably Poland, Finland, Moldova, Sweden, Georgia and quite probably Kazakhstan and some of the other former parts of the Russian Empire. Such policies are openly discussed by Russian politicians, in the Russian media and on Russian television. Even if he is not successful in Ukraine he may attack other countries. He has gambled everything on at least being able to claim military victory somewhere. His supporters expect that of him and of Russia.

Again there are parallels with the Second World War. Once Hitler came to power he gradually abolished any checks and balances on his power, and made bigger and bigger military gambles. As long as he achieved some kind of victory his support grew. The adulation became ever greater. Had the international community prevented his re-occupation of the Rhine-lands, or the bombing of Guernica, or the Anschluss with Austria, or his marching into the Sudeten-lands of Czechoslovakia, he might never have invaded Poland, France, Russia and the whole Second World War might have been prevented.

Had Putin's military aggression in Chechnya, or Syria, or his 2014 annexation of Crimea and the Donbas been less successful we might not be in the situation that we are in today. Putin has now, as Hitler once had, a lot of fanatical supporters. They crave glory to reinforce their sense of racial superiority. Every defeat erodes that sense of superiority and support for the leader who personifies it. The sooner Putin can be emphatically defeated the quicker a peaceful path of progress can be restored. Such a victory would also weaken the malign forces of division, racism and inequality that permeate global politics and society.

It is also worth noting that big steps forward often emerge from times of crisis and upheaval. In the depths of the Second World War the Beverage Report was written and hopes of a better future inspired the country. The Attlee government founded the NHS and made many sweeping reforms that improved the lives of the ordinary British people in those immediate post-war years. It is imperative the world now unites to beat Putin, and to usher in a more socially just and ecologically sustainable future.

Kamil Galeev is a young academic and activist of Russian background, who has studied in Russia, China, Scotland and the USA. He is a prolific user of Twitter, posting long threads of well-informed commentary on Russian history, sociology, military strength and he makes very well-informed speculations about the possible future for Putin, Russia and the region. Galeev sees three possible scenarios for the future of Russia.[142]

Firstly, there is what he calls the 'North Korea/Donbas' scenario. Putin stays in power after being brought to a standstill in the Ukraine. Russia, from the outside would look increasingly like North Korea, a fanatically militarized state where the Russian population remained utterly impoverished and a tiny elite lives an isolated bunkered existence. How the Donbas has been run since coming under Russian occupation in 2014 shows how the country would be run. There are essentially no economic opportunities beyond working for the state, no rule of law or democratic freedoms. A system of state sponsored local warlords runs the region. For most people the system has resulted in brutality and impoverishment. In order to keep the people in check the county has to be kept under a constant war footing.

Kamil Galeev's second scenario is what he refers to as 'Imperial Reboot'. Here he paints a picture of Putin being ousted and being replaced with another leader. The Russian Federation would remain a vast, centralized, brutal system led by another ethnic Russian chauvinist; with the many ethnic minorities of the former Russian empire still very much second-class citizens. Galeev portrays even Navalny fitting into this racist and imperialistic mode, even though he is more liberal and democratic than any of the likely alternatives.

Kamil Galeev's third, and preferred, scenario is what he calls 'National Divorce'. Here he sees the Russian Federation breaking apart. There are dozens of regions of the country where a great diversity of ethnic groups bears little allegiance to Moscow. Many of these regions pay a heavy price for Russian domination and have an urge for greater self-governance. Galeev sees the Russian Far East as perhaps the weakest link, but many areas might erupt in rebellion against Moscow's domination. It could be a peaceful or a violent process of disaggregation. Let us hope for a peaceful process.

As I have noted earlier, nearly all very large countries are poorly governed. Size does seem to matter. Networks of smaller states co-operating together seems to be a better model, exemplified by the five Nordic countries, and increasingly also the three Baltic Republics. These networks are also linked into other networks of co-operation such as the Sustainable Cities Platform and the European Union. It is with these multi-layered collaborative networks that the best existing forms of governance are to be found. In this book I have extended this understanding into my imagined network of 39,000 networked communities. That global network of communities is a theoretical proposition. In reality a process of ever-increasing devolution may chaotically evolve. Russia is in effect an imperial power, as Britain was in the nineteenth century. Many of its colonized communities yearn for greater freedom from Russian domination and the fragmentation of the Russian Federation may happen more quickly than most Westerners understand.

Ukraine has made impressive strides forward. It is now better governed and less corrupt than it has ever been. Since the 2014 Russian occupation of Crimea and Donbas Ukraine has organised and modernized its defensive capabilities, modelled in part on the Finnish system of total defence. Small groups of infantry equipped with shoulder-launched missile systems like the Javelin and NLAW have been able to stop whole columns of Russian tanks in the wooded countryside of northern Ukraine. Driving the Russians out of the more open countryside of Donbas and Crimea, and overcoming Russian air and sea domination will require other weapon systems.

While Putin is in power, and after what he has unleashed on Ukraine, I see no prospect for a negotiated peace. Not until after a Ukrainian victory. It is imperative that the rest of the world supports Ukraine. Leadership is coming from the Baltic Republics, Poland, the Czech and Slovak republics, Slovenia and a number of other smaller European states, and Finland and Sweden will now probably join NATO to help protect themselves from likely Russian aggression. Key EU people such as Ursula Von der Leyen and Guy Verhofstadt have expressed strong support for a tough line against Russian aggression. Sadly, France and Germany have to some extent been dragging their feet both in terms of sanctions and supplying weapons. The USA is proving useful in the background and Joe Biden has unequivocally referred to the Russian action as genocide. The position of the UK is confused and chaotic.

The UK: Conflicted, Confused and Chaotic

The UK is at a very strange point in its history. Scores of our MPs, over many years, have accepted Russian money and free holidays and other perks from Putin and his network of oligarchs and KGB/FSB operatives. Our prime minister, Boris Johnson, makes frequent visits to his friend Evgeny Lebedev, ennobling him despite his strong and known KGB/FSB connections. David Smith, writing in the Guardian, shows how the KGB cultivated Donald Trump as an asset from 1977.[143] Boris Johnson may well have been similarly cultivated from as early as his student days. Shallow narcissistic characters like Johnson and Trump must have been easy targets for the KGB.

Investigative journalists and authors such as Carole Cadwalladr, Sam Bright, Oliver Bullough, Adam Bienkov, Peter Jukes, Tom Burgis, Peter Geoghegan and many others are doing their best to expose the complex ways in which British politics has been corrupted by Russian money. The 'Led by Donkeys' organization has made a number of You Tube videos based on its findings. Precious little of this is reported in our mainstream media.

Russian money has flowed into London, buying property, making charitable donations, paying school fees to posh public schools, making political donations, funding so-called think-tanks and offering free holidays and perks to the British elite. The Conservative Friends of Russia, Vote Leave, The Taxpayers Alliance, The Global Warming Policy Foundation and a number of other shady organizations, many based at 55 Tufton Street, often with overlapping membership, have had a deeply corrosive impact on British politics. The Quintessentially Group boasted about how it could arrange meetings for wealthy Russian clients.

The ordinary Russian people, like the ordinary American people and the ordinary British people, are rapidly becoming more impoverished. Wealth is flowing upwards toward an ever more closely linked global oligarchy. Our Chancellor of the Exchequer, Rishi Sunak, and his wife, Akshata Murthy, are typical of this global elite, profiting from dodgy dealing in Russia, paying no tax by being 'non-doms', with American Green card residency to further aid their ability to escape taxation. During the Covid pandemic the ten richest billionaires have doubled their wealth, while the incomes of 99% of humanity fell.[144]

Brexit, Covid, rising fuel prices and most of all the ideologies of market fundamentalism, are contributing to the mass impoverishment of the average British person. Brexit always was a policy of oligarchic power. The oligarchs' client politicians make use of populist rhetoric, and their ownership of so much of our media convinced many poor people to vote to leave the EU. P&O's fire and re-hire would have been illegal under EU law, and naturally French or Belgian workers were not treated like this, yet the British dockers' union had advocated members vote to leave the EU. In the 2019 general election the Conservatives won a landslide, buoyed up by the very votes of those they intended to impoverish. Under this government corporate profits for oil, gas and water companies have been surging and shareholder dividends generous, while fuel prices skyrocketed and sewage flowed into our rivers.

It does not have to be like this. The populations of the Nordic region and many other places are not being similarly exposed to processes of impoverishment and pollution. Oligarchic capitalism in the UK could in theory be reigned in. It is all a question of political will. The Conservative agenda, should they remain in power, looks set to further impoverish the vast majority of the British population. The health service has been going through a process of creeping privatization for some years, and this looks set to speed up towards a full-scale privatized system based on the American one. This would spell utter disaster for most British people.

Channel Four television is up for privatization. No doubt the government hope and expect it to be bought by some billionaire oligarch whose worldview matches their own. But what if it was bought by someone, or some community of people, whose views were very different?

What if a number of groups and organizations got together and bought the station? Let us imagine a network of dozens, or hundreds, of individuals, groups and organizations doing such a thing. Oxfam, Friends of the Earth, Greenpeace, Christian Aid, Extinction Rebellion, The School Strikes Movement, Avaaz and many other social and environmental justice organizations would be a great part of the team. They might want to team up with many investigative journalists, or cleantech companies, or institutional investors with a social conscience. As Jeremy Leggett's Scottish Rewilding is attempting mass ownership of land, might someone or some group initiate a mass ownership model of running a TV station?

One of the reasons why we are so poorly governed, are being systematically impoverished and exposed to insane levels of pollution is because better ways of managing things are simply not discussed in our media. I would love to see a television station that was dedicated to exploring in detail how we could have a more ecologically sustainable and socially just future, starting now. The channel would not bother showing much in the way of sport, drama, films or the arts, which are covered by the BBC and others. What we are sorely in need of is something different.

Imagine a television station where George Monbiot and Carole Cadwalladr presented a nightly programme investigating government corruption and malpractice. Where Mark Z Jacobson and Chris Goodall presented programmes on renewable energy and the cleantech revolution, where Greta Thunberg introduced programmes on climate change and Mark E Thomas looked at processes of mass impoverishment and how to reverse it. There could be lots of programmes about land use; Patricia Kombo exploring improving food security in Africa, Byron Kominek on what role agrivoltaics might play, Chris Packham and George Young presenting how rewilding and food production could best be combined. It would be great to see Kamil Galeev, Anders Ostlund and Lesia Vasylenko presenting programmes to the UK public on the unfolding situation in Ukraine, Russia and the surrounding lands. Underpinning everything would be how we could best transition from turbo charged consumer driven capitalism towards something very different, and very much better, and how taxation and the government legislative programme could be used to help bring about that change.

Onwards to a Better Future

This book has been my attempt to help shift the Overton Window and help introduce ideas that may help us all achieve a better future. We could disempower Putin and bring peace to the Ukraine by abruptly halting all fossil fuel imports into the UK and EU. Energy efficiency measures and renewable energy co-operatives could be a vehicle to lower carbon emissions, while also weakening Putin and helping ordinary people get warmer homes and lower bills. The impact of this book may well be tiny. Having a television channel to develop these ideas could have a transformative impact on society. My absolute dream job would be presenting a television programme introducing the ideas, technologies and projects that I have been enthusing about in this book and the incredible people behind them.

Amory Lovins and colleagues from the Rocky Mountain Institute have just published a paper titled 'From Deep Crisis, Profound Change'[145] looking at how Putin's war is pushing up fossil fuel prices, just as renewable energy costs continue to fall. Now is exactly the right time to make a very rapid switch from fossil fuels to renewables. This is the fastest way to end Putin's war in Ukraine and to act on the climate crisis. Such a global transition will create many millions of jobs. Lovins is mainly looking at technological change, and a little about behavioural change. If one added some of the ideas presented in this book on, for example, changes to taxation, then progress could be even faster and more widely beneficial.

Throughout this book I have endeavoured to show how a multitude of interlinked problems could be addressed simultaneously. Wendell Berry developed a concept called 'solving for pattern' whereby a solution to one problem did not inadvertently create other problems but instead was aimed at mitigating numerous pressing problems. He originally applied the concept to land management, but since it has been applied to many other areas including system change.

The UK media is full of stories about refugees and economic migrants seeking to cross the English Channel in small rubber boats. These stories are often specifically designed to foster fear, suspicion and hatred. As I write these final paragraphs of the postscript Priti Patel has just announced a plan to send asylum seekers to Rwanda for 'processing'. This has disturbing echoes of how the Nazis spoke of 'processing' the Jews.

Other stories deal with issues from climate change to homelessness as separate stories, competing for our attention and prioritization. I have endeavoured to show how solutions could be interlinked. Reframing these three stories I have attempted to show how by creating meaningful work for absolutely everyone who wanted to participate we could provide work, solving problems like forced migration, climate change and homelessness. I have endeavoured to show the big structural changes needed, and the tiny efforts we all can make, and how they are interlinked.

Bringing peace to Ukraine, ending mass impoverishment and very rapidly reducing carbon emissions could all be achieved in a matter of months. The only thing lacking has been political will. Globally there are already many millions of activists pressing for these kinds of changes. We need many millions more. Tipping points happen, and things can very suddenly change, for better, or for worse. Please, in whatever way you can, help push for a more ecologically sustainable and socially just future. Together, we can and we must bring about a political tipping point, before the global climate reaches its own horrific tipping points.

Contact: Questions and Updates

If you have questions about this book you can contact me by e-mail:
richardgpgsblog@hotmail.com

I will continue blogging and updating the ideas behind this book.
I will also post the text, references and links of the book on to the
website. Do please sign up to get blogs sent directly to your inbox.

www.richardpriestley.co.uk

Endnotes

1   https://www.imf.org/en/Topics/climate-change/energy-subsidies

2   https://e360.yale.edu/digest/fossil-fuels-received-5-9-trillion-in-subsidies-in-2020-report-finds

3   https://www.sipri.org/media/press-release/2020/global-military-expenditure-sees-largest-annual-increase-decade-says-sipri-reaching-1917-billion

4   https://www.atlanticcouncil.org/global-qe-tracker/

5   https://www.oxfam.org/en/press-releases/carbon-emissions-richest-1-percent-more-double-emissions-poorest-half-humanity

6   https://www.theguardian.com/environment/2021/sep/14/four-in-10-young-people-fear-having-children-due-to-climate-crisis

7   https://www.psychologytoday.com/gb/blog/out-the-darkness/201907/pathocracy

8   https://www.visionofhumanity.org/wp-content/uploads/2020/10/GPI_2020_web.pdf

9   Anu Partanen 'The Nordic Theory of Everything' Duckworth 2018

10   https://wellbeingeconomy.org/about

11   https://wellbeingeconomy.org/wego

12   https://en.wikipedia.org/wiki/High_Ambition_Coalition

13   https://en.wikipedia.org/wiki/2019_Herefordshire_Council_election

[14]   To safeguard future generations, we must learn how to be better
       ancestors | Environmental activism | The Guardian

[15]   https://www.worldometers.info/world-population/

[16]   https://www.richardpriestley.co.uk/carbon-emissions-billionaires-
       the-bbc/

[17]   https://www.theguardian.com/environment/2008/mar/23/
       ethicalliving.lifeandhealth4

[18]   The term 'Market Fundamentalist' was coined by George Soros in
       1998, and the best analysis of what the term means is to be found in
       Chapter 4 of '99%: Mass Impoverishment and How We Can End
       It' by Mark E Thomas, Apollo 2019

[19]   Despair, drug use and suicide have also become more common and
       are also driven by extreme inequality and are major factors in this
       trend of falling American life expectancy, and the mishandling of
       the Covid pandemic has only exacerbated the situation.

[20]   https://theconversation.com/private-planes-mansions-and-
       superyachts-what-gives-billionaires-like-musk-and-abramovich-
       such-a-massive-carbon-footprint-152514

[21]   https://www.oxfam.org/en/press-releases/carbon-emissions-
       richest-1-percent-more-double-emissions-poorest-half-humanity

[22]   https://transformdrugs.org/blog/drug-decriminalisation-in-
       portugal-setting-the-record-straight

[23]   https://www.ted.com/talks/johann_hari_everything_you_
       think_you_know_about_addiction_is_wrong?language=en

[24]   https://ourworldindata.org/energy

[25]   https://news.mit.edu/2018/explaining-dropping-solar-cost-1120

[26]   https://www.carbonbrief.org/solar-is-now-cheapest-electricity-in-

history-confirms-iea

27 https://100percentrenewableuk.org/invitation-to-tender-for-a-100-per-cent-renewable-uk-model-scenario

28 https://www.offshorewind.biz/2021/12/07/siemens-gamesa-looks-into-replacing-offshore-power-cables-with-pipes/

29 https://www.euractiv.com/section/energy-environment/news/hydrogen-trade-hopes-boosted-by-australia-germany-deal/

30 https://www.smh.com.au/business/small-business/electrify-everything-cannon-brookes-calls-for-east-west-solar-cables-to-power-australia-20200907-p55t59.html

31 https://gravitricity.com/projects/

32 https://ourworldindata.org/cheap-renewables-growth

33 My blog is searchable, and there is a lot about various solar technologies https://www.richardpriestley.co.uk/

34 https://www.richardpriestley.co.uk/morocco-pioneering-solar/

35 https://renewablesnow.com/news/barilla-to-test-new-solar-thermal-technology-at-italian-pasta-plant-717098/

36 https://www.solarthermalworld.org/news/lisbon-solar-cooling-system-office-complex and see also https://www.richardpriestley.co.uk/air-conditioning-refrigeration/

37 https://www.coagrivoltaic.org/

38 https://www.pv-magazine.com/2020/07/23/special-solar-panels-for-agrivoltaics/

39 https://chinadialogue.net/en/energy/with-coal-off-the-menu-can-china-export-its-renewables-model/ and there is a great video https://www.youtube.com/watch?v=_Uv-Yw9MZFE

40 https://www.richardpriestley.co.uk/?s=wind

41 https://www.richardpriestley.co.uk/floating-wind/

42 https://www.richardpriestley.co.uk/floating-wind-comes-to-ireland/

43 https://en.wikipedia.org/wiki/Balbina_Dam

44 https://en.wikipedia.org/wiki/Dinorwig_Power_Station

45 https://www.statkraft.com/about-statkraft/where-we-operate/norway/kvilldal-hydropower-plant/

46 http://www.tidallagoonpower.com/about/

47 https://www.emec.org.uk/

48 https://www.richardpriestley.co.uk/big-heat-pumps/

49 http://fintrydt.org.uk/

50 https://chamberlainhighburytrust.co.uk/2021/04/28/joseph-chamberlains-time-in-birmingham/

51 https://www.irena.org/-/media/Files/IRENA/Agency/Publication/2021/Apr/IRENA_-RE_Capacity_Highlights_2021.pdf?la=en&hash=1E133689564BC40C2392E85026F71A0D7A9C0B91#:~:text=Total%20renewable%20capacity%20in%202019,538%20GW%20(%2B0.06%25).

52 https://slowways.uk/

53 https://billmckibben.substack.com/p/the-happiest-number-ive-heard-in

54 https://www.richardpriestley.co.uk/big-heat-pumps/

55 https://ourworldindata.org/sanitation

56 https://ourworldindata.org/water-access

57 https://ourworldindata.org/energy-access

58 https://www.bbc.co.uk/iplayer/episode/moo11wn5/panorama-
    the-electric-car-revolution-winners-and-losers

59 https://www.cam.ac.uk/stories/landsparing

60 https://www.carbonbrief.org/guest-post-how-enhanced-
    weathering-could-slow-climate-change-and-boost-crop-yields

61 https://blog.whiteoakpastures.com/hubfs/WOP-LCA-
    Quantis-2019.pdf

62 https://foodfoundation.org.uk/covid_19/monitoring-imports-
    of-fruit-and-veg-into-the-uk/#:~:text=When%20it%20comes%20
    to%20fruit,%5BHort%20Stats%2C2018%5D.

63 https://www.nationalgeographic.com/
    magazine/2017/09/holland-agriculture-sustainable-
    farming/#:~:text=U.S.&text=The%20tiny%20Netherlands%20
    has%20become,land%20available%20to%20other%20countries.

64 As above

65 https://www.lowcarbonfarming.co.uk/ingham-bury-st-edmunds/

66 https://www.gothamgreens.com/

67 Special solar panels for agrivoltaics – pv magazine International
    (pv-magazine.com)

68 https://www.pv-magazine.com/2020/09/03/giant-agrivoltaic-
    project-in-china/

69 https://www.richardpriestley.co.uk/qatar-project/ and https://
    www.saharaforestproject.com/qatar/ and https://www.
    saharaforestproject.com/revegetation/

70 https://www.saharaforestproject.com/jordanian-women-trained-

on-modern-agricultural-technology-at-sfp/

71 https://www.theguardian.com/sustainable-business/2017/jan/23/aquaculture-bivalves-oysters-factory-farming-environment

72 https://offshoreshellfish.com/about-us/ and https://www.positive.news/environment/food/farms-of-the-future-mussel-farming-given-a-new-lifeline/

73 https://en.wikipedia.org/wiki/Edible_seaweed

74 https://www.seawatersolutions.org/news-articles/a-saline-solution-for...

75 https://theconversation.com/eating-insects-has-long-made-sense-in-africa-the-world-must-catch-up-70419#:~:text=The%20dominant%20insect%20eating%20countries,termites%2C%20crickets%20and%20palm%20weevils.

76 See Merlin Sheldrake 'Entangled Life: how fungi make our worlds, change our minds and shape our futures' Penguin Random House, 2020

77 https://www.theguardian.com/environment/2019/jun/29/plan-to-sell-50m-meals-electricity-water-air-solar-foods and https://www.theguardian.com/commentisfree/2020/jan/08/lab-grown-food-destroy-farming-save-planet

78 https://treesforlife.org.uk/about-us/narrative/

79 https://www.bunloit.com/team and https://www.bunloit.com/

80 https://www.highlandsrewilding.co.uk/blog

81 http://www.fao.org/state-of-forests/en/

82 https://earth.org/how-costa-rica-reversed-deforestation/

83 https://oceanservice.noaa.gov/facts/oceanwater.html#:~:text=About%2097%20percent%20of%20Earth's,in%20

glaciers%20and%20ice%20caps.

[84]   https://www.nationalgeographic.com/news/2017/10/niue-chile-marine-parks-ocean-conservation-environment/

[85]   https://www.nytimes.com/2019/09/24/world/europe/italy-tuscany-fishing-art.html

[86]   https://en.wikipedia.org/wiki/Great_Stink

[87]   http://www.bbc.co.uk/earth/story/20151111-how-the-river-thames-was-brought-back-from-the-dead

[88]   http://webcam.wyeuskfoundation.org/problems/pollution.php and https://www.theguardian.com/environment/2020/jun/20/its-like-pea-soup-poultry-farms-turn-wye-into-wildlife-death-trap

[89]   https://www.thefold.org.uk/carefarm/

[90]   https://www.sekem.com/en/about/history/

[91]   https://whiteoakpastures.com/pages/about-us

[92]   https://www.haygrove.com/our-history/

[93]   https://www.canonfromecourt.org.uk/about-us/

[94]   Dennis Hardy 'Alternative communities in nineteenth century England' 1979 Longman, London

[95]   https://globalsolaratlas.info/map?c=39.774769,-3.515625,2&s=-16.636192,50.976563&m=site

[96]   https://news.masdar.ae/en/news/2021/01/25/11/29/eci-collaborates-with-masdar-to-increase-uaes-renewable-energy-infrastructure-development

[97]   https://www.theguardian.com/world/2019/dec/18/how-water-is-helping-to-end-the-first-climate-change-war

98 There is a lovely 6 minute introduction to the Machakos Miracle on YouTube https://www.youtube.com/watch?v=EI_yRArxw14

99 https://www.greatgreenwall.org/results

100 https://www.greatgreenwall.org/about-great-green-wall

101 https://www.greenthesinai.com/home

102 https://www.richardpriestley.co.uk/my-technology-of-the-year/

103 https://en.wikipedia.org/wiki/Qanat

104 https://en.wikipedia.org/wiki/Ouarzazate_Solar_Power_Station

105 https://www.azelio.com/media/press-releases/2020/azelio-inaugurates-its-renewable-energy-storage-at-noor-ouarzazate-solar-complex-in-morocc

106 https://www.richardpriestley.co.uk/big-solar-in-egypt-dubai/

107 Lehman, David, ed. (2006). The Oxford Book of American Poetry. Oxford University Press. p. 184

108 https://fuelcellsworks.com/news/namibia-announces-9-4-billion-green-hydrogen-project/

109 https://www.richardpriestley.co.uk/repowering-port-augusta/

110 https://en.wikipedia.org/wiki/Toshka_Lakes

111 https://en.wikipedia.org/wiki/Neom

112 https://www.h2bulletin.com/saudi-neom-blueprint-hydrogen-city/

113 https://www.statkraft.com/newsroom/news-and-stories/archive/2021/social-responsibility-life-giving-water-in-peru/

114 https://www.bbc.com/future/article/20211115-how-morocco-led-the-world-on-clean-solar-energy

115 https://www.h2greensteel.com/fossil-free-steel-plant

116 Three links to explain this paragraph: https://www.liquidwind.
se/news/liquidwind-partners-with-orsted-to-produce-green-
electro-fuel-in-large-scale-emethanol-project-in-sweden
and https://smartcitysweden.com/best-practice/305/
energy-heat-and-steam-generated-from-99-renewable-
fuels/#:~:text=H%C3%B6rneborgsverket%2C%20a%20
biofuel%2Dbased%20cogeneration,of%20%C3%96vik%20E-
nergi's%20energy%20production.&text=Cogeneration%20
plants%2C%20also%20known%20as,all%20energy%20into%20
useful%20energy. And https://www.rechargenews.com/energy-
transition/shipping-giant-maersk-to-become-major-green-
hydrogen-consumer-as-it-embraces-methanol-fuel/2-1-1143147

117 https://www.zerosun.se/page?id=24550

118 See Anu Partanen 'The Nordic Theory of Everything', Duckworth
2018

119 https://bicycledutch.wordpress.com/2016/01/05/motorway-
removed-to-bring-back-original-water/#:~:text=It%20took%20
another%20ten%20years,restore%20the%20historic%20city%20
moat.&text=The%20North%20part%20of%20the,lot%20for%20
almost%2030%20years.&text=This%20part%20of%20the%20
city%20moat%20returned%20in%20January%202002.

120 https://www.treehugger.com/netherlands-kids-take-bicycle-bus-
school-4857337

121 https://www.theguardian.com/world/2019/aug/25/cargo-bikes-
berlin-four-wheels-bad-transport

122 https://www.coastfutura.org/

123 Tragically the Dearman company is currently in administration.
The technology is brilliant, but they got into financial difficulties. It

is a company that needs rescuing.

[124] https://www.cprehereford shire.org.uk/issues-were-dealing-with/river-wye-pollution/

[125] https://www.pressandjournal.co.uk/fp/opinion/3348760/climate-change-shaming-washing-dishes-corporations-opinion/

[126] https://www.theguardian.com/science/2021/jul/19/billionaires-space-tourism-environment-emissions

[127] https://theconversation.com/private-planes-mansions-and-superyachts-what-gives-billionaires-like-musk-and-abramovich-such-a-massive-carbon-footprint-152514

[128] https://fridaysforfuture.org/september24/

[129] https://caminotocop.com/about-us/

[130] The talk, but not the questions and answers, is available via a link on the events page of my blog. https://www.richardpriestley.co.uk/events/

[131] https://www.richardpriestley.co.uk/greens-electoral-breakthrough/

[132] https://www.richardpriestley.co.uk/the-green-party-reflections-hopes/

[133] https://en.wikipedia.org/wiki/Next_United_Kingdom_general_election

[134] https://en.wikipedia.org/wiki/North_Herefordshire_(UK_Parliament_constituency)

[135] https://sonomotors.com/en/sion/

[136] https://www.theguardian.com/lifeandstyle/2008/apr/16/recipes.foodanddrink

[137] https://wakelyns.co.uk/bakery/wholegrains/

[138] https://wakelyns.co.uk/populations/

[139] https://hodmedods.co.uk/collections/pulses-grains-seeds

[140] https://www.herefordpedicabs.com/

[141] Martin Sixsmith: 'Russia: A 1,000 year chronicle of the wild east' Random House 2011. It was wonderfully abridged for Radio 4, and now is very much due for an update.

[142] https://twitter.com/kamilkazani/status/1509991598595223554

[143] https://www.theguardian.com/us-news/2021/jan/29/trump-russia-asset-claims-former-kgb-spy-new-book

[144] https://www.oxfam.org/en/press-releases/ten-richest-men-double-their-fortunes-pandemic-while-incomes-99-percent-humanity

[145] https://rmi.org/insight/from-deep-crisis-profound-change/